W9-DAC-229

Astronomer's Pocket Field Guide

For other titles published in this series, go to
www.springer.com/series/7814

Rony De Laet

The Casual Sky
Observer's Guide

Stargazing with Binoculars
and Small Telescopes

Springer

Rony De Laet
Renaat Woutershof 18
3460 Bekkevoort, Belgium
rodelaet@yahoo.com

ISBN 978-1-4614-0594-8   e-ISBN 978-1-4614-0595-5
DOI 10.1007/978-1-4614-0595-5
Springer New York Dordrecht Heidelberg London

Library of Congress Control Number: 2011935990

Printed on acid-free paper

Springer is part of Springer Science+Business Media (www.springer.com)

Preface

Congratulations on your acquisition of this copy of *The Casual Sky Observer's Guide*. It is your starting point for a great journey through time and space.

On our journey, we are not alone. Each clear night thousands of amateur astronomers all over the world are making the same voyage with you. Many generations have gone before us. Our ancestors watched the stars with great intensity. The night sky was their timekeeper, calendar, and compass. The night sky announced the changing of the seasons. It dictated when to plant and when to harvest. The stars seemed to rule about life and death on Earth. Our ancestors saw patterns in the sky. They imagined things of importance, such as hunters and quarry, kings and queens, evil and good. They told great stories about the constellations in the night sky. We carry with us the oral traditions of thousands of years of stargazing.

When we see the twinkle of the stars, we look at the same stars our ancestors looked at. The stars are our origin and our destination. Today, society seems to ignore this legacy. During our modern day to day worries, we don't take the time to care about the rising and setting of a celestial object. We're simply too busy with our work and family. As if we don't want to know where we came from nor where we're heading.

In the year 2009, a year before the writing of this book, we celebrated the International Year of Astronomy (IYA 2009). It was a time to reflect on the 400th anniversary of the first telescopic observations recorded by the Italian astronomer Galileo Galilei. Galileo discovered craters and mountains on the Moon. He was the first to note that Jupiter has its own moons, that Venus shows phases like our Moon, and that there are many more stars than the naked eye can see. Proud as he was, he wrote down his recordings and drawings in what we could call a first scientific treatise based on telescopic observations.[1] In 1610 he published his work *Siderius Nuncius, or the Starry Messenger*. Galileo's conclusions incurred the hostility of his contemporaries. Society was not ready to accept his discoveries, nor was the Catholic Church pleased with another world model. Freedom of thought was not encouraged in 1609. Although the telescope was eagerly accepted for its military use, its astronomical application was considered rather offensive. Galileo only tried to understand the universe as he saw it in his telescope. He was put on trial in 1633, for suspicion of heresy. He was ordered to abandon all ideas contradictory to the Holy Scripture.

---

[1] The English astronomer Thomas Harriot is credited as the first observer to draw an astronomical object with use of a telescope. He made a map of the moon, several month before Galileo did. Unfortunately, Thomas Harriot never published his work.

Publication of any of his astronomical work was strictly forbidden. Galileo eventually lived out his final days under house arrest.

Four hundred years later, we live in somewhat better times when it comes to scientific investigation. In a large part of the world, freedom of thought is widely accepted. Science has made great progress. And our views on the universe are totally different than 400 years ago. Humankind has slipped away from the center of the antique universe towards the crusty surface of an insignificantly little planet near a desolate dwarf star in the backyard of a mediocre galaxy. The telescopic instruments have been greatly improved. Galileo had to build his own telescopes. They were rather crude tools, and the images that they produced were blurred by the imperfections in the glass that was used for the ground and polished lenses. Today instruments are much improved and refined. For the price of a good meal, you can buy a decent pair of binoculars, which optical quality-wise will outperform any of Galileo's instruments by great measure.

The best news is that a pair of binoculars is all you need to join in our casual exploration of the Milky Way and the universe. Chances are good that you already own a pair of binoculars or that you know someone who has a pair you can borrow. It might surprise you, but many of the destinations that we will visit are even visible with the naked eye. So for the time being, there is absolutely no need to buy a large telescope.

Our exploration of the universe is divided in twelve monthly chapters. Each month, different wonders of the deep sky will be covered. As such, you can use the book throughout the year, as the seasons come and go. When you have a clear evening and some spare time to spend under the stars, just start with reading the appropriate chapter. Each object is accompanied by a finder chart and a drawing. The finder chart will help you to locate the desired object. The drawing will show you what to look for and what to expect to see with your binoculars. These drawings are the author's personal pencil sketches. They show my perception of the object. I deliberately did not use photographs, because they do not represent the night sky in the same way that our eyes do. Throughout this guide you will hone your observing skills, and when you understand the true nature of the deep-sky objects, you'll learn to see with your mind's eye as well. When you're not familiar with the constellations, you can use the all-sky star maps at the beginning of the book. These maps will help you to locate the constellations of interest.

Before you dive into one of the deep-sky chapters, do read the preceding chapters first. These will help you to prepare for your journey with additional equipment and techniques in the best possible way. *The Casual Sky Observer's Guide* is all you need to start observing from your own backyard as well as from a fine holiday spot.

Broaden your horizon and have yourself a great trip among the stars!

Bekkevoort, Belgium   Rony De Laet

Acknowledgements

The writing of this book would not have been possible without the help and support of the following people.

First of all I would like to thank the true stars in my life, my wife Birgit and my children Hanne and Michiel for their great support during the writing of this book.

I want to express my deepest admiration for the late Carl Sagan, whose exciting television series, *Cosmos,* triggered my fascination for the stars when I was a child. I could not have imagined then that I would ever observe and sketch the wonderful cosmos, about which he could narrate in such a magnificent manner.

A special thank you goes to my beloved parents, who bought me my first telescope when they understood my ambitious interest in exploring the mind-staggering depths of the universe.

I also want to thank Stephen James O'Meara, whose book, *The Messier Objects,* opened my eyes. With his book, O'Meara taught me how to become a better observer through sketching. Though I never had the privilege to meet him personally, I often felt like I was looking over Stephen's shoulder through his telescope. The marvelous drawings of Stephen James O'Meara inspired me to make my own series of sketches of the universe.

A very special thank you goes to Richard Handy, a very experienced observer and a talented sketcher. His sketches of the Moon are truly works of art. Richard has always encouraged me to keep on sketching. He is (co-)author of two astronomy books and the inspiring creator of the website 'Astronomy Sketch of The Day' (www.asod.info).

I want to thank Pierre Chevalier, the developer of the software *Sky Charts* (Cartes du Ciel). *Sky Charts* is a complete planetarium program. It is used by many amateur astronomers to prepare their observations. Pierre Chevalier was so kind to make *Sky Charts* available for free. This versatile charting program allowed me to draw the finder charts for this book.

I am grateful to Bill Tschumy, the creator of the three-dimensional galactic atlas, 'Where is M13.' His innovative application helps us to visualize the true location of deep-sky objects in space. With the use of 'Where is M13,' I was able to create the unique galaxy views for the book. Bill Tschumy's software is available for free.

I also want to thank John Watson and Maury Solomon for giving me the opportunity to write this book.

Finally, I want to thank my good friend Thomas McCague for his outstanding support during the writing of this book. Tom has been an amateur astronomer and telescope maker for more than 40 years. He is not only a passionate observer and a gifted sketcher, but he also hosts public star viewings at the G. Jack Bradley Observation Deck of the Moraine Valley Community College in Illinois. Tom has been my sounding board from the very initiation of this book. Having an experienced amateur observer ready to evaluate the illustrations and the texts was of inestimable value to me. I fear to count the sleepless nights he spent on reviewing the drafts of the text. Tom and I are both passionate star gazers with devotion for sketching. Our conversations have helped shaped this book in its final appearance. Tom, it was a privilege to be able to rely on your generous assistance during the development of the manuscript. Thank you, my friend.

About the Author

Rony De Laet, a Belgian national who holds a Master of Science degree in Industrial Science (Chemistry), has been an enthusiastic amateur astronomer since his teens. He has had articles published (in Dutch) in the monthly Flemish VVS Astronomical Magazine, and his special talent is in producing photo-real computer drawings of the night sky. His work was on exhibition at the International Astronomical Sketching Exhibition, called "In the Footsteps of Galileo," at the Blackrock Castle Observatory in Ireland (from February until May 2009) and then later at Birr Castle, the historic site of Rosse's 1845 72-inch telescope (now restored and open for visitors).

Contents

NAVIGATING THE SEA OF STARS

Fig. 1.1 The Western sky at dusk in April, as seen from a hill in Belgium. When you know your way between the stars, you'll recognize the constellations of Gemini and Auriga. If you're not familiar with the constellations, see Fig. 1.4, where you can find Gemini and Auriga. The brightest object, in the lower right of the picture, is not a star but our neighbor planet, Venus

R. De Laet, *The Casual Sky Observer's Guide*, Astronomer's Pocket Field Guide, DOI 10.1007/978-1-4614-0595-5_1, © Springer Science+Business Media, LLC 2012

Before we embark on our journey through the universe, we need to have a look at our preparation. Like on a real journey, certain steps are primordial for the success of the trip. In this chapter, we'll have a look at our roadmaps, destinations, luggage, and the kind of vehicle we want to use to navigate the stars (Fig. 1.1). All this may sound complicated, but it should not. With a few adequate measures, our voyage into deep space is quite straightforward. Let's first have a look at the wonders of the universe visible to us on a clear night.

Objects in the Deep Sky

On our journey through space we will encounter various types of objects. What we call the deep sky is the space in our universe beyond the Solar System. The Solar System includes the Sun and all the planets, the Moon, the comets, and the asteroids. Every star other than our Sun lies in the deep sky. In fact all the individual stars that you can see with the naked eye belong to our Milky Way, which is also part of the deep sky.

Throughout the book we will use the light-year (ly) as the measuring rod for the distances in our universe. The light-year is the distance that light travels in a vacuum in 1 year. One ly is equal to 9.5 trillion km, or 6 trillion miles. Sirius, the brightest star in the sky, lies 8.6 ly away from us. The light that we receive from Sirius has traveled for 8.6 years to arrive at Earth. When we look at Sirius, we also look into the past, because we see how Sirius was 8.6 years ago. The Sun lies at a distance of 8 light-minutes. It takes 8 min and 19 s for the light of the Sun to reach Earth. Hence we see the Sun how it was 8 min ago. In space we cannot travel a distance without traveling back in time.

Stars are distant Suns. They are massive, hot, incandescent balls of mainly hydrogen and helium, with surface temperature ranging from 3,600 to 50,000 K. We can see them because of the light that they emit, as a by-product of the nuclear fusion in their core.

Here follows the various types of objects that we will encounter in the deep sky.

Constellations

A constellation is an imaginary pattern of prominent stars. Our ancestors believed that these stars belonged together. Today we know that most patterns are formed by stars that have no astrophysical relationship with each other. These stars lie light-years apart. In modern astronomy, a constellation is a well defined area in the sky. The sky is divided into 88 official constellations.

Variable Stars

These are stars of which the apparent brightness changes in time. This can happen because the luminosity of the star itself varies, or because some of the light of the star is blocked from reaching us.

Multiple Stars

Many stars appear to have an optical companion when observed in binoculars or telescopes. Some stars just happen to lie in the same line of sight without being physically related. Other optical double or triple stars are truly bound by gravity. When the components of a multiple star system are very close together, we cannot separate them visually. They appear as a single point of light. Astronomers, however, can use a spectroscope to analyze the light of the star system. The spectroscopic fingerprint of the starlight then reveals the presence of multiple stars. Such star systems are called spectroscopic multiple stars.

Star Clusters

Large groups of stars that are bound together by gravity are called star clusters. There are two types of star clusters. Globular clusters are massive spherical collections of stars, held firmly together by gravity. Most globular clusters are far away from Earth and thus too faint to be noticed with the naked eye. A few of them are visible in binoculars. Open clusters (also called galactic clusters) are loosely bound groups of stars that are born in roughly the same place and time. These clusters don't exhibit enough gravitational attraction to keep their stars together indefinitely. Many open clusters are easily visible with the naked eye on a clear night. Our Sun, which is a solitary star now, was probably born in such a galactic cluster.

Nebulae

Nebulae are the alpha and the omega of the stars. Diffuse nebulae are vast clouds of dark and cold interstellar matter; dust and hydrogen. It is from these clouds that stars and planets form. They become emission nebulae when they emit light, because they are ionized by the strong radiation of young, hot nearby stars. On our journey, we will visit various emission nebulae.

Dark nebulae have no powerful stars nearby and don't emit light. We can still see them as dark patches against a brighter background. The Great Rift in the summer Milky Way is a conglomeration of such dark nebulae. Planetary nebulae are the ejected outer shells of a star in its death throes. These nebulae have nothing in common with planets whatsoever. When these objects were discovered in the eighteenth century, they were mistakenly categorized as giant planets. Planetary nebulae play an important role in chemical evolution of the universe. It is through these nebulae that the stars return an important part of their content back to the interstellar medium. Since most of the stars produce planetary nebulae at the end of their life, we could expect that there is an abundance of these planetaries to be found in the interstellar medium. However, this is in not the case. We can only see a few planetary nebulae with our binoculars because they are rather small and faint, and relatively short-lived.

The last type of nebulae that we'll discuss are supernova remnants. These are the debris clouds produced by the violent explosion of a very massive star at the end of its life. These explosions are inappropriately called supernovae. Nova means new star. Once more, this type of nebulae returns star stuff back to the interstellar medium.

Galaxies

Simply put, galaxies are the giant collections of all of the above-mentioned objects, held together by gravity. Each object is in orbit around the center of the galaxy, like the planets of our Solar System are in orbit around the Sun. All the stars that you see from your own backyard with the naked eye, including our Sun, are part of the Milky Way, our own galaxy. During summer nights, the Milky Way is visible with the naked eye as a pale band of light arching around the entire sky. When you point your binoculars at this pale band, you will be able to resolve its light into a myriad of faint stars. Our Milky Way, which is only a modest galaxy by universal standards, counts a total of 200–400 billion stars!

Our journey through the universe will bring us near several other galaxies as well. There are three main types of galaxies. Spiral galaxies, like our Milky Way, are large, flattened pinwheels with prominent spiral arms and with a spherical central hub. Elliptical galaxies have the shape of a flattened sphere. Irregular galaxies show no regular shape. There are probably 170 billion galaxies in the universe as we know it. Most galaxies are gathered in clusters and super clusters. However, don't let these numbers convince you that space is a crowded place. Galaxies are better compared with distant islands in a very empty ocean.

Navigating the Night Sky

The hardest part of deep-sky observing is locating the desired object in the sky. We are familiar with the movement of the Sun during the course of the day. The Sun rises in the east, culminates in the south, and sets in the west. The sky as we see it changes throughout the course of the day and night due to the rotation of our planet. To make things even more complex, our planet also orbits the Sun. Every season, a different part of the night sky is presented to us.

The Seasonal Sky Maps

To help you find your way among the stars, we've included four pairs of sky maps (Figs. 1.3–1.10). Each pair represents the complete evening sky after dusk for one of the four seasons for observers in mid-northern latitudes. We can consider these sky maps as our nighttime window to the universe. These sky maps represent the location of the different constellations together with the glowing band of the Milky Way for a particular season. The curved edges of these maps correspond with your horizon. The center of each pair shows the zenith. This is the point directly overhead. Before you start using the maps, you should be aware of the direction you're looking in. The direction of sunset is west. If you look to the west, north is to your right. South is to your left, and east is behind your back. Once you know these cardinal points, you can start using the all sky maps. The curved edges of these maps are labeled with the cardinal points in yellow.

Suppose you want to see the constellations in the south. Take the appropriate pair of maps for the season you're in, and rotate the book until the label "south" is right side up. Hold the book in front of you and compare the stars in the book with the ones in the southern sky. That's all there is to it.

Why four pairs of sky maps? The answer is very simple. We can only see stars when the Sun is below the horizon; otherwise the sky is too bright. We all call this the nighttime. Our view of the universe is thus restricted by the glare of the Sun. When we look up into space during the night, we look in the opposite direction of where the Sun is. At night, our window to the universe lies away from the Sun.

Earth orbits the Sun in a slow but constant movement. We call the period of time it takes to complete a full orbit around the Sun a year. This means that every 6 months, we are at opposite sides of our orbit around the Sun. It also means that every 6 months, our nighttime window to space is on opposite directions as well. Therefore during the course of 1 year, we see constellations come and go through our nighttime window. And after a year, the whole process starts over again (Fig. 1.2).

You can verify that our planet is in orbit with a simple test. If you point a telescope at a bright star in the south, you will see that the star drifts out of view. This is caused by the rotation of Earth. Now imagine that you leave the telescope in this fixed position. The next night, the bright star will pass

Fig. 1.2 The orbit of Earth around the Sun. We can only see the stars on the side opposite of the Sun. The *yellow* stars represent the summer constellations. The *red* stars represent the autumn constellations. The *blue* stars represent the winter constellations. The *green* stars represent the spring constellations

again through the field of view of your telescope. But if you timed the passage, you would see that the star is 3 min and 56 s earlier than the night before. This is caused by the slow movement of Earth around the Sun.

Compared to the Sun, Earth rotates once in exactly 24 h. Compared to the stars, Earth rotates in 23 h and 56 min. This apparent extra rotation speed among the stars comes from the one revolution around the Sun in a year. Divide 24 h by 365 (the days in a year) and you have your 3 min and 56 s.

You can follow the course of the constellations throughout the year on the successive seasonal maps also. The most significant consequence is that some constellations are only visible during a particular time of the year, when we are in the appropriate part of Earth's orbit around the Sun. So we will have to be humble and patient if the constellation of our interest lies on the opposite side of Earth's orbit. It makes no sense to search for Orion during summer nights, or to plan a visit to the Archer on winter evenings. The same is true for our view of the Milky Way, as is shown in the seasonal maps. Spring evenings in particular are not favorable for an occasional stroll along the brighter star fields of our home galaxy. However, our universe harbors so many beautiful treasures that every month of the year is interesting enough to go deep-sky hunting.

Did you also notice on the seasonal sky maps the thin line with the label "ecliptic"? The ecliptic represents the path that the Sun and the planets (and also the Moon) follow in the sky during the year. It is in fact the projection of the Sun among the stars, due to our orbit around the Sun. The ecliptic can be imagined as a grand circle, with us in the center of it. At high noon around the 23rd of June, the Sun is high in the sky. It is at the highest point of the ecliptic. Around the 23rd of December, the Sun arrives at the lowest point of the ecliptic. These two dates, just like any other two dates that are 6 months apart, indicate opposite positions on this circle which we call the ecliptic. Remember that the seasonal sky maps are meant to be used when night falls and the Sun has set. Therefore, at night, you see the ecliptic on the opposite side of where the Sun is.

Because the ecliptic is a grand circle, the following rule applies: when the ecliptic is high at noon (June 23), it will be low at midnight and vice versa (see also Figs. 1.9 and 1.5). That explains why planets and the Moon shine high in the sky on winter nights and low above the southern horizon on summer nights. The foundation for the curious path of the ecliptic is the inclination of the rotation axis of Earth compared to the plane of its orbit around the Sun. This also explains why we have the seasons.

Occasionally it will happen that you run into a bright object that is not represented on the sky maps. Several bright objects, such as planets, so-called shooting stars, or satellites are not included on the maps because they don't have fixed positions between the stars. Satellites may reflect sunlight in such a manner that a temporary bright flare is seen in the sky. Some satellites appear to move in the sky, while others appear fixed between the stars.

Shooting stars, or falling stars, are in fact not stars at all but meteoroids. They are mostly sand-sized particles from our Solar System that enter Earth's atmosphere. The visible streak of light that occurs when a meteoroid enters Earth's atmosphere and disintegrates is what we call a meteor. Meteor showers occur when Earth passes through the trail of debris left by a comet. Planets are not included on the sky maps because they wander along the ecliptic. Mercury, Venus, Mars, Jupiter and Saturn are so bright that they can be seen even with the naked eye! If you see a bright "star" along the ecliptic and it is not plotted on the maps, then you have discovered a planet. See the last chapter for resources on planet ephemerides.

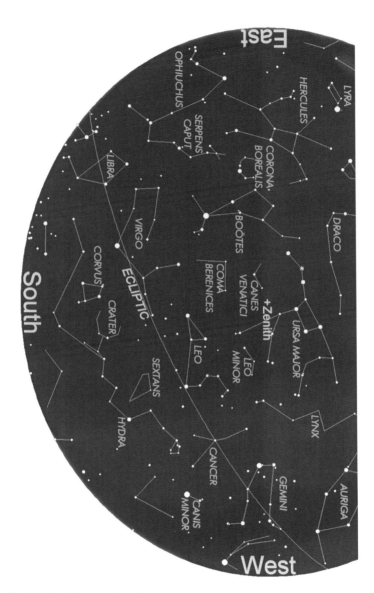

Fig. 1.3 Southern spring sky

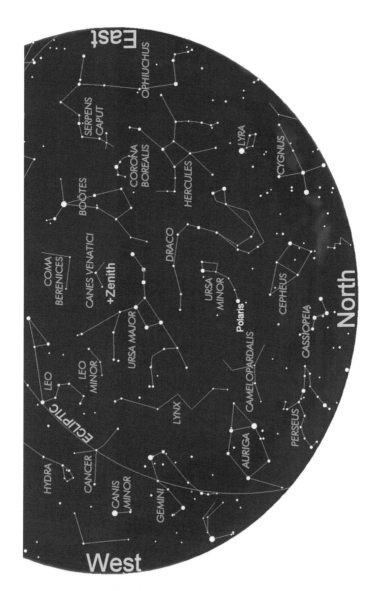

Fig. 1.4 Northern spring sky

Fig. 1.5 Southern summer sky

Fig. 1.6 Northern summer sky

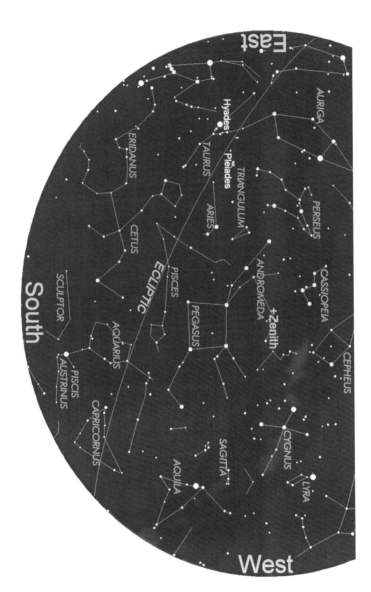

Fig. 1.7 Southern autumn sky

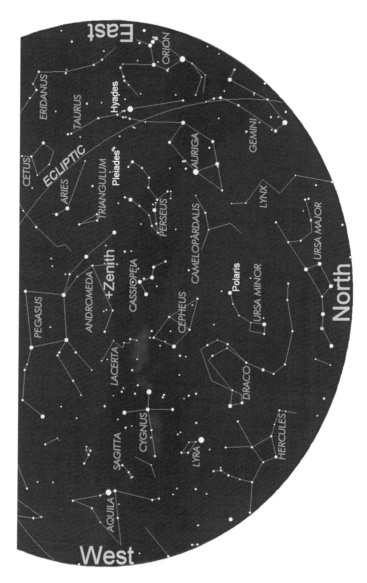

Fig. 1.8 Northern autumn sky

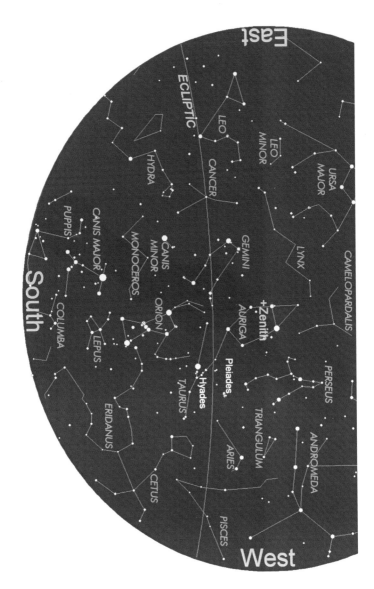

Fig. 1.9 Southern winter sky

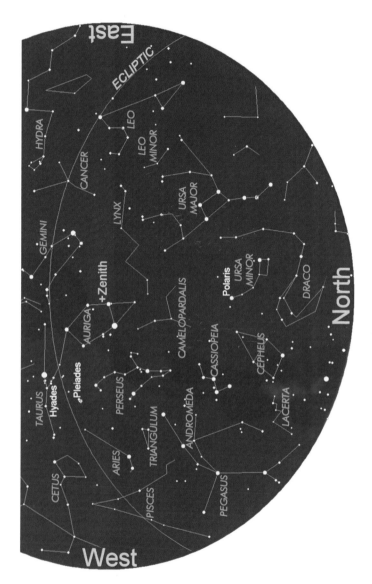

Fig. 1.10 Northern winter sky

Polaris and the Pointer Stars

These charts might intimidate you at first, certainly if you're not used to navigating between constellations. There are a few tricks to keep things simple. When you examine the northern sky maps of the different seasons, you'll notice that they have a few things in common. Firstly, they all show a star labeled Polaris. That's the famous North Star. Did you notice that Polaris is always marked on the same spot of the different maps? That's right, Polaris does not move! Now, be careful, the North Star is NOT the brightest star in the sky. It just happens to lie very close to the north celestial pole, which is the projection of the axis of Earth in the sky.

Secondly, all stars appear to move in circles around Polaris. All heavenly objects seem to rotate counterclockwise around the north celestial pole, because of our planet's daily rotation. An object at the north celestial pole does not move. Now, we can take advantage of this phenomenon. Polaris shows us where north is (it's always straight above the northern horizon), and Polaris's angular height above the horizon corresponds always with the latitude of the observer. If you observe from latitude 40° North, Polaris appears 40° above the northern horizon. If you would observe from the North Pole, Polaris would be at the zenith, 90° above the horizon. An observer at Earth's equator cannot see Polaris, as it lies on the northern horizon. Thus Polaris is an important reference point when navigating the skies. It should be one of the first stars to go after when starting an observation session. Because Polaris is not a bright star, we need some kind of a signpost to find it. Unfortunately, there are no signposts in the sky, so we will have to make our own. The best technique is to "star hop" from one star to another until we arrive at the desired location in the sky. But first we need to learn some of the constellations.

The right position to start with is the Big Dipper, which is an asterism within the constellation Ursa Major, the Great Bear. The Big Dipper is a conspicuous far northern group of bright stars and is visible all night throughout the year. The Big Dipper is shown in Fig. 1.11.

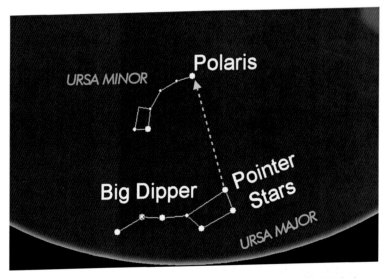

Fig. 1.11 The Big Dipper is an asterism that can be used to find Polaris

The three leftmost stars are the handle, the four remaining stars form the bowl. In some countries, it's also called the Plough or the Great Wagon. Now pay attention to the two rightmost stars of the bowl. These two stars are called the pointer stars, as they point towards Polaris if you follow the line between them to the north. If you take a look at Fig. 1.8, you'll see the Big Dipper right above the yellow label for north.

Now compare Fig. 1.8 with Figs. 1.4, 1.6 and 1.10. Do you see how the Big Dipper moves around Polaris throughout the seasons, as do all the other stars on the maps as well? This is a very important point: all stars seem to rotate counterclockwise around the North Star during the course of the night, but also throughout the seasons. So no matter where the Big Dipper is, its pointer stars always show us Polaris *and* the true north. We have just discovered a celestial compass that's always available when the nights are clear!

Now, let's summarize what the seasonal maps show us about the Big Dipper. During spring evenings, the Big Dipper is high overhead, with the handle pointing to the east. During summer evenings, the Big Dipper is lower in the northwest, with the handle pointing up. During autumn evenings, the Big Dipper hovers above the northern horizon, its handle pointing to the west. If you don't have a clear view to the northern horizon, it is possible that the Big Dipper is hidden behind nearby trees or buildings during the autumn evenings. And during the spring evenings, the Big Dipper rises in the northeast, its handle pointing down. Luckily you don't have to know these positions by heart if you have the seasonal maps with you. The Big Dipper will be a stepping stone to find other constellations or objects as well, so practice locating this useful asterism during every clear night under the stars.

We call the constellation of Ursae Major (the Great Bear) a circumpolar constellation. Circumpolar objects never set or rise. They can be seen on every clear night throughout the year. Which stars are circumpolar depends on the latitude of the observer. Our example counts for mid-northern latitudes, because the sky maps are plotted for a latitude of 45°N. For an observer at the North Pole, all constellations are circumpolar. An observer at the equator will see no circumpolar stars. You can see which constellations are circumpolar for mid-northern latitudes on the sky maps. The constellations and stars that are not circumpolar are called seasonal. An observer at the equator will only see seasonal constellations, while his or her colleague at the North Pole sees none. While circumpolar constellations can be seen on every clear night, seasonal constellations are only visible during a given time of the year.

Directions and Angular Distances in the Sky

Polaris indicates an important location in the northern sky: the north celestial pole. This is the direction in which the rotation axis of Earth points. There is of course also a south celestial pole, which lies on the opposite site of where Polaris is. We, observers north of the equator, cannot see the south celestial pole, because that point lies behind our southern horizon at exactly 180° away from Polaris.

Because the celestial poles have an angular distance of 180°, any star that lies 90° away from the north celestial pole also lies at 90° of the south celestial pole. Such a star lies at the location we call the celestial equator. The celestial equator is a great circle equally separated in angular distance from the celestial poles. Be aware that exactly half of the celestial equator is above our horizon. The celestial equator is easy to locate: point your left arm at Polaris and hold your right arm at a straight angle to your left arm. Your right arm points towards the celestial equator. Now turn around while keeping your left arm pointed at Polaris and your right arm at a straight angle. If you can complete a full circle, your right arm has described the great circle we call the celestial equator.

The celestial equator cuts our horizon exactly in the east point and in the west point. A star located at the celestial equator rises exactly in the east point, culminates in the south at an angular height of 90° minus the latitude of our observing site (if our latitude is 40° then the culmination height is 50°), and sets exactly in the west point. So you see how this grand circle, the celestial equator, is different from the other grand circle, the ecliptic. The celestial equator maintains a fixed inclination in the sky, whereas the ecliptic's orientation depends on what part of the sky we look at.

Now that we have learned how to locate the North Star, Polaris, we can have a look at the celestial directions. We already know that the Sun rises in the east and sets in the west. The Sun also rotates around Polaris, just like all the other stars in the sky. For that reason, we call the direction in which the celestial objects drift, west. East is then in the opposite direction. Be aware of the fact that these directions in the sky do NOT correspond with the directions we use on Earth! Take another look at Fig. 1.8, and see where the Big Dipper is. Assume that you would be outside on a clear autumn evening, to watch the movement of the Big Dipper around Polaris. You would see that the Big Dipper moves in a counterclockwise direction around Polaris. From your point of view, the Big Dipper moves to the east of your location, but we still call it a celestial westward drift. *All stars move to the west*, even the ones that we see under Polaris.

How about north in the sky? That's very simple: *north is where Polaris is*. If you have an object in the field of view of your binoculars, the edge closest to Polaris is the northern edge of your field of view. This sounds logical at first, but in the field, it can appear very awkward.

Let's assume that you watch from an observing site, with a clear horizon in all directions, how the moon rises in the east. The northern side of the Moon is then on your left, the side where Polaris is. If you waited long enough to see the Moon set in the west, the northern side of the Moon is – this time to your right side. A good trick to use with binoculars or a telescope is the following: the direction of drift of the stars in your field of view (fov) is west. When you tilt your 'scope towards Polaris, that side is north. All the finder charts and sketches all the way through this book have north up and west to the right. When you look up objects in the field, you should always try to hold the finder charts so that the top of the chart points towards Polaris. That way, the chart is oriented in the same way as the real sky is.

Now that we know how to use celestial directions, we can take the next step – measuring distances in the sky. We can describe distances between objects with angles, measured in degrees. This method is very straightforward. The angular distance between the horizon and the point straight overhead, the zenith, is 90°. A half circle, from south to zenith, to north measures 180°. If you point your hand towards the horizon, and you turn right to complete a full circle, you've covered 360°.

Frequently we use our hands and fingers to measure angular distances. Just hold your hand at arm's length in front of the sky. Your wide open hand covers almost 20° from thumb to little finger. Your fist (with the thumb inside) should cover about 10°, while the three middle fingers value 5°. Your little finger covers just about 1°. Why is this useful? Well, the Big Dipper covers 25° from the tip of the handle to the pointer stars. When you think you found this asterism, try to cover it with you wide open hand. The Big Dipper should be slightly larger than your wide open hand. The distance between the two pointer stars is 5°, the same amount of sky covered by three fingers at arm's length. The width of the bowl of the Big Dipper measures 10°. You can cover that with one fist. The distance from the pointer stars to Polaris is 28°. This means that you can cover the line from the pointer stars to the North Star with your wide open hand, or three fists.

Smaller angular distances than one degree are frequently used as well. One degree is divided into 60 arcminutes, or 60'. One arcminute equals 60 arcseconds, or 60". Two familiar objects are measured in arcminutes. Both the disk of the Sun and the full Moon cover about 30', or ½°. The middle star of the handle of the Big Dipper is actually a double star, called Alcor and Mizar. Their separation is 14 arcminutes, about half the size of the lunar disk. A healthy human eye has a resolution of nearly 1 arcminute. Therefore it is possible to separate Alcor and Mizar with the unaided eye. Give it a try, next time you locate the Big Dipper.

What else can we measure with angular distances? Two important concepts: the apparent size of a deep-sky object, and the field of view of your binoculars or telescope. The apparent size of a deep-sky object equals the angular distance from one edge of the object to the other edge. The apparent size of the Andromeda Galaxy, for example, measures 3° × 1°. The true fov of your binoculars is also measured in angular distance. Most regular binoculars have a true fov ranging from 5° to 7°.

Why is it important to know the fov of your binoculars? Because you can use the fov as a measuring rod to estimate angular distances and sizes. Once you know the fov of your own pair of bino's, you can calculate approximately what portion of the fov will be covered by the apparent size of a deep-sky object. For example, if your binoculars have a true fov of 6°, you can hold both the pointer stars in one single field of view. For that same pair of binoculars, the angular distance from the center of the fov to the edge of the field equals 3°. The disk of a full Moon would fit 12 times in that same fov (see Fig. 1.12). The finder charts in this book all have a binocular fov circle of 6°. It will be important to compare that value to the fov of your own binoculars. Telescopes usually show a much smaller true fov than binoculars, because of the higher magnification they work at. That's why they are equipped with a finder scope, a little scope with a large fov, to point the telescope at the desired location.

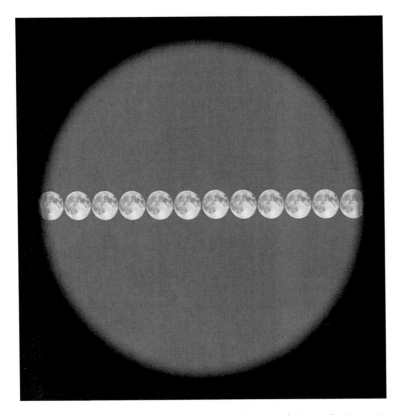

Fig. 1.12 This sketch represents the field of view of the author's 8×56 binoculars. They have a true fov of nearly 6°. The full Moon has an apparent size of 30'. The disk of the Moon fits 12 times in the fov of my binoculars

Stars and their Properties

Star Names and Constellations

The brightest stars are labeled with lower case Greek letters or a number identification. The Greek letters (followed by the Latin genitive form of the constellation) were introduced by the German astronomer Johann

Table 1.1 The Greek alphabet

α	Alpha	ι	Iota	ρ	Rho
β	Beta	κ	Kappa	σ	Sigma
γ	Gamma	λ	Lambda	τ	Tau
δ	Delta	μ	Mu	υ	Upsilon
ε	Epsilon	ν	Nu	φ	Phi
ζ	Zeta	ξ	Xi	χ	Chi
η	Eta	ο	Omicron	ψ	Psi
θ	Theta	π	Pi	ω	Omega

Bayer in 1603. Some stars have multiple designations. Polaris, for instance, is also labeled α (alpha) *Ursae Minoris*, or in short α *UMi*. The brightest star in the sky is named Sirius, but it is also labeled as α (alpha) *Canis Majoris*, or α *CMa*. Yes, this hobby of ours brings along some odd practices to master. Because most of us are not used to reading in Greek or Latin, we've included a table for both the Greek alphabet and the constellations in Latin. Tables 1.1 and 1.2 will become very handy when you start reading the next chapters.

Mid-northern observers can only see a part of all the 88 constellations. The rest of the constellations remain hidden behind the observer's horizon. Although Table 1.2 might look weird, dull and old-fashioned from a modern human's perspective, it was a milestone in the history of our species. With this list of 88 constellations, humankind had mapped the universe to its own liking. It was a brave venture to encompass the whole universe. The list does contain some very old constellations – VERY old. We inherited most of the constellations from the ancient Greeks. But the Greeks had not invented all of ancient constellations themselves. About 30 constellations were imported by the Greeks from older civilizations!

Recent research, based on reports from excavations in Mesopotamia (modern-day Iraq), has revealed that these ancient constellation names are Sumerian. This may not ring a bell to you, but Sumer is the earliest known

Table 1.2 The 88 official constellations in Latin

Abbreviation	Constellation	Latin genitive	Meaning
And	Andromeda	Andromedae	Damsel In Distress
Ant	Antlia	Antliae	Air Pump
Aps	Apus	Apodis	Bird-Of-Paradise
Aqr	Aquarius	Aquarii	Water-Bearer
Aql	Aquila	Aquilae	Eagle
Ara	Ara	Arae	Altar
Ari	Aries	Arietis	Ram
Aur	Auriga	Aurigae	Charioteer
Boö	Boötes	Boötis	Herdsman
Cae	Caelum	Caeli	Chisel
Cam	Camelopardalis	Camelopardalis	Giraffe
Cnc	Cancer	Cancri	Crab
CVn	Canes Venatici	CanumVenaticorum	Hunting Dogs
CMa	Canis Major	Canis Majoris	Great Dog
CMi	Canis Minor	Canis Minoris	Little Dog
Cap	Capricornus	Capricorni	Sea-Goat
Car	Carina	Carinae	Keel
Cas	Cassiopeia	Cassiopeiae	The Queen
Cen	Centaurus	Centauri	Centaur
Cep	Cepheus	Cephei	The King
Cet	Cetus	Ceti	Sea-Monster
Cha	Chamaeleon	Chamaeleontis	Chamaeleon
Cir	Circinus	Circini	Drafting Compass
Col	Columba	Columba	Dove
Com	Coma Berenices	Comae Berenices	Berenice's Hair
CrA	Corona Australis	Coronae Australis	Southern Crown
CrB	Corona Borealis	Coronae Borealis	Northern Crown
Crv	Corvus	Corvi	Crow
Crt	Crater	Crateris	Cup
Cru	Crux	Crucis	Cross
Cyg	Cygnus	Cygni	Swan
Del	Delphinus	Delphini	Dolphin
Dor	Dorado	Doradus	Dolphinfish
Dra	Draco	Draconis	Dragon
Equ	Equuleus	Equulei	Foal
Eri	Eridanus	Eridani	River
For	Fornax	Fornacis	Furnace
Gem	Gemini	Geminorum	Twins

(continued)

Table 1.2 (continued)

Abbreviation	Constellation	Latin genitive	Meaning
Gru	Grus	Gruis	Crane
Her	Hercules	Herculis	Hercules
Hor	Horologium	Horologii	Pendulum Clock
Hya	Hydra	Hydrae	Water-Serpent
Hyi	Hydrus	Hydri	Water-Serpent
Ind	Indus	Indi	Indian
Lac	Lacerta	Lacertae	Lizard
Leo	Leo	Leonis	Lion
LMi	Leo Minor	Leonis Minoris	Little Lion
Lep	Lepus	Leporis	Hare
Lib	Libra	Librae	Balance
Lup	Lupus	Lupi	Wolf
Lyn	Lynx	Lyncis	Lynx
Lyr	Lyra	Lyrae	Lyre
Men	Mensa	Mensae	Table Mountain
Mic	Microscopium	Microscopii	Microscope
Mon	Monoceros	Monocerotis	Unicorn
Mus	Musca	Muscae	Fly
Nor	Norma	Normae	Level
Oct	Octans	Octantis	Mariner's Octant
Oph	Ophiuchus	Ophiuchi	Serpent-Bearer
Ori	Orion	Orionis	Hunter
Pav	Pavo	Pavonis	Peacock
Peg	Pegasus	Pegasi	Winged Horse
Per	Perseus	Persei	Hero
Phe	Phoenix	Phoenicis	Phoenix
Pic	Pictor	Pictoris	Painter's Easel
Psc	Pisces	Piscium	Fishes
PsA	Piscis Austrinus	Piscis Austrini	Southern Fish
Pup	Puppis	Puppis	Stern
Pyx	Pyxis	Pyxidis	Mariner's Compass
Ret	Reticulum	Reticuli	Eyepiece Graticule
Sge	Sagitta	Sagittae	Arrow
Sgr	Sagittarius	Sagittarii	Archer
Sco	Scorpius	Scorpii	Scorpion
Scl	Sculptor	Sculptoris	Sculptor

(continued)

Table 1.2 (continued)

Abbreviation	Constellation	Latin genitive	Meaning
Sct	Scutum	Scuti	Shield of Sobieski
Ser	Serpens	Serpentis	Serpent
Sex	Sextans	Sextantis	Astronomical Sextant
Tau	Taurus	Tauri	Bull
Tel	Telescopium	Telescopii	Telescope
Tri	Triangulum	Trianguli	Triangle
TrA	Triangulum Australe	Trianguli Australis	Southern Triangle
Tuc	Tucana	Tucanae	Toucan
UMa	Ursa Major	Ursae Majoris	Great Bear
UMi	Ursa Minor	Ursa Minoris	Little Bear
Vel	Vela	Velorum	Sails
Vir	Virgo	Virginis	Virgin
Vol	Volans	Volantis	Flying Fish
Vul	Vulpecula	Vulpeculae	Little Fox

civilization. The first Sumerian settlements began in the middle of the sixth millennium B.C. Some of today's constellations, like Aquila, Auriga, Gemini, Orion and Sagittarius, have their primitive Sumerian counterparts carved in 6,000-year-old stone cylinder seals. The Sumerians handed down their culture and knowledge, and with it their constellations, to younger civilizations.

One of the most prominent constellations in the sky is Orion. Orion has some kind of a mystic appeal. Our ancestors noticed Orion's extraordinary appearance as well. Some civilizations saw a hunter in Orion. The Sumerians, on the other hand, called it "the True Shepherd of Heaven". Constellations are really universal in time and place (on Earth). Orion has not changed much in appearance during the last six millennia. Today we see it just like our ancestors saw it 6,000 years ago. And it is still quite exciting to see the constellation of Orion emerging out of the dusky sky on a clear spring evening (Fig. 1.13).

Fig. 1.13 The Hunter and the Bull are setting in the evening sky. The constellations of Orion and Taurus are a legacy from our ancestors of more than 6,000 years ago. Today we still see them just like the Sumerians did in their era (except for the light pollution!). Use Fig. 1.9 to find these constellations on the seasonal map

Star Brightness

A star's brightness is measured in magnitudes (mag). It's an old method, introduced by the ancient Greeks. The stars were divided into six categories. The brightest stars belonged to the class of first magnitude, the faintest visible to the naked eye were called sixth magnitude stars. The Greeks didn't have telescopes in their era, so they were not aware of the fact that there are fainter stars than the naked eye can see. The lower the magnitude number, the brighter the star. A first magnitude star is 2.5 times brighter than a second magnitude star. A second magnitude star is 2.5 times brighter than a 3rd magnitude star, and so on. Thus a first magnitude star is $2.5 \times 2.5 \times 2.5$ times brighter than a 4th magnitude star. A first magnitude star is 100 times brighter than a sixth magnitude star. The Greek system is still with us, albeit with a few modern refinements. Nowadays, the brightness of a star is measured up to 0.001 of a full magnitude. The scale is also extended beyond first and sixth magnitude. The brightest stars are in fact too bright to belong to the class of the first magnitude. These stars even have zero or negative magnitudes. Remember that the lower the number, the brighter the star is. Sirius, the brightest star, has a magnitude of −1.4. Venus can shine with a maximum brightness of −4.4! The full Moon measures mag −12.6. We can even put our Sun on the magnitude scale: mag −26.8. If you use a pair of binoculars or a telescope, you will be able to see fainter stars than those of the sixth magnitude. The magnitude scale is infinite.

Our perception of the brightness of a star, as seen from Earth, is called the apparent magnitude. It does not tell us anything about the true luminosity of a star. Stars lie at different distances from us. Sirius appears bright with its apparent magnitude of −1.4 because it lies only 8.6 light-years (ly) away. Polaris (the North Star) on the other hand has a dimmer apparent magnitude of 1.97, but it lies at 430 ly from us. Astronomers use absolute magnitude to measure the true luminosity of the stars. The absolute magnitude of a star is the apparent magnitude if the star were placed at a imaginary distance of 32.6 ly from us.

The absolute magnitude gives us a better understanding of the true luminosity of the objects in the sky. The absolute magnitude of Sirius is 1.4. The absolute magnitude of Polaris is −3.6! Thus Polaris is in fact the brighter one. Does the Sun have an absolute magnitude? Yes, we can imagine our Sun placed at a distance of 32.6 ly. The Sun's absolute magnitude is only 4.8. It's in fact dimmer than Sirius.

For extended objects, such as clusters, nebulae or galaxies, astronomers use integrated magnitude, which is the brightness the object would have if all of its light was focused in a single point of light. The integrated magnitude (both apparent and absolute) allows us to compare the luminosities of extended objects. The integrated absolute magnitude of our home galaxy, the Milky Way (if all of its light were concentrated in a single point at 32.6 ly distance) is a whopping −20.5. That's an impressive number, but the integrated absolute magnitude of the Andromeda Galaxy is −21.7!

Star Types

Stars come in different sizes and colors. The colors of the stars that we see in our binoculars or telescopes depend on the surface temperatures of the observed stars. You can compare the colors of the stars with the colors in the flame of a candle. Blue corresponds with the highest temperature. Next come blue-white, yellowish white, yellow and orange. The coolest stars appear red.

Astronomers have divided the stars in different spectral classes, depending on their surface temperature. The main spectral classes, in order of descending surface temperature, are: O, B, A, F, G, K, and M. O-stars are the hottest, while M-stars are the coolest. Each class is divided in ten subclasses ranging from 0 to 9. Our Sun is a G2 star. Sirius is an A1 star.

Stars are formed from the clouds of interstellar dark matter, like the Great Rift that divides the summer Milky Way (see Fig. 2.9). The spectral type of a newborn star depends on its initial mass. Heavy O-stars radiate a lot more energy than the lighter M-stars. These massive stars use up their fuel very quickly. An O5 star with a surface temperature of over 25,000 K, typically has 40 solar masses (40 times the mass of our Sun). It will burn up in about five million years and end its existence in a deadly supernova explosion. A G2 star with one solar mass (like our Sun) has a surface temperature of only 6,000 K. Such a star has an estimated lifetime of ten billion years. The "coolest" M-stars with a mass of only 0.1 solar mass have a very low fuel consumption. They can keep their candle burning for hundreds of billions of years. In our galaxy there are more lighter than heavier stars to be found. Interstellar clouds produce more lighter stars than heavier stars, and the lighter stars have the longest lifespan. The massive O- and B-stars in fact are rare in our galaxy (Table 1.3).

Table 1.3 Spectral classes of stars

Type	Surface temp (Kelvin)	Color	Example of a subclass	Absolute magnitude	Typical mass (solar mass)	Estimated lifetime (years)
O	>25,000	Blue	O5	-6.8	40	5 million
B	25,000–11,000	Blue	B0	-4	18	10 million
			B5	-1.5	6.5	60 million
A	11,000–7,600	Blue-white	A0	0.3	3.2	500 million
			A5	1.5	2.1	2 billion
F	7,600–6,000	Yellow-white	F0	2.8	1.7	5 billion
			F5	3.8	1.3	9 billion
G	6,000–4,500	Yellow	G2	4.8	1	10 billion
K	5,100–3,200	Orange	K0	6.5	0.78	20–30 billion
			K5	7.5	0.69	
M	<3,200	Orange-red	M0	9	0.47	>100 billion?

The Art of Observing

How Dark Is Your Sky?

It sounds logical that the night sky appears dark and transparent. Unfortunately this is most often not the case. Many of us live in a populated area where light pollution rules the night. Light pollution occurs when the light from cities and industries is scattered in the atmosphere, rendering it less transparent. If you live in or near a city, your sky will probably not be dark enough for deep-sky observing. The night sky will appear gray instead of dark and only the brightest stars will show up. To make things even worse, your eyes will not get the chance to dark-adapt. There is only one solution: choose a more convenient (remote) observing site. A small pair of binoculars at a dark observing site will offer more observing pleasure than a large 'scope in the city. If you're not able to move out, try to make your backyard as dark as possible by blocking out any stray light from nearby lights. It helps to observe from under a shroud if necessary. The darkness of the sky can be expressed by the naked-eye limiting magnitude, or nelm. A nelm of 4.8 means that the faintest stars that can be discerned by the naked eye near the zenith are of magnitude 4.8. The higher the nelm, the better. There are only a few places in Europe were the nelm is 6.5 or better. One of these places is an hour and a half driving from the author's home. It is time well spent to look out for a convenient observing location. The best sites are high and dry. The higher you get, the thinner the air to look through. Desert regions or mountains often guarantee a very high nelm.

But how faint can the human eye see? In the dedicated literature you'll often find the mag 6.5 limit for the maximum nelm. But that's an average limit. The truth is that no one can predict your personal maximum nelm. It'll depend on your eyesight, your experience with stargazing, and your observing site. Under ideal circumstances, the healthy human eye is capable of detecting stars up to mag 8!

The nelm of your observing site is of great importance for your observations, and you should take note of it when you observe. The nights with the higher nelm will allow you to see the deep-sky objects better.

How Clear Is Your Sky?

As stated before, the transparency of the sky as seen from our backyard varies from night to night and even from hour to hour. Clouds, humidity, air pollution, light pollution and moonlight are big factors. On average there are about 60 clear nights in many parts of Europe and North America. A third of these clear nights come without proper announcement. It happens that at the end of a cloudy day, the cloud cover breaks up and disappears in just a few hours right after sunset. Keep an eye on the latest weather reports and on the real-time, web-based satellite images. The colors of the western sky right after sunset reveal how promising the night will be. If the dusky sky shows all the colors of the rainbow, with plenty of greenish tints, as in Fig. 1.1, then the atmosphere is very transparent. If the air contains a lot of smog or humidity (see Fig. 1.14), the sky colors at dusk will appear washed out. That is a bad omen for the deep-sky observer. The glare of the Moon, especially during the 14 days around full

Fig. 1.14 When the air is polluted with smog, sunset appears very *red* in color. This is a bad omen for the deep-sky observer

Moon (that is, from first quarter to last quarter), will also wash out the stars from your sky. If possible, try to observe on a moonless night, because these are the darkest ones.

But when the sky is clear and dark, the atmosphere can still be very turbulent. You can measure the turbulence by the twinkling of the stars. Stars near the horizon are more affected by turbulence, because we look through a thicker layer of air there than when we look at the stars near the zenith.

Astronomers use the term "seeing" for measuring the steadiness of the atmosphere. If the air is calm, we say that the "seeing" is very good. Seeing can be influenced by strong winds or the jet stream but also by local conditions, such as the neighbors' warm roof or a chimney over which you want to look at the stars. If conditions allow, try to observe from a grass field, instead of from a concrete pavement. Avoid nearby buildings as well. They slowly release their heat throughout the night, inducing air currents that interfere with your observations.

The seeing becomes even more important when you want to observe with high magnification. If you observe with an ordinary pair of 8 × 40 binoculars, the seeing won't affect your observing that much. But it's still important to take note of the seeing for future reference. Why is that? When the seeing is excellent, the stars will appear tack sharp in your binoculars or telescope. Fainter stars will be easier to see. Your telescope or binoculars seem to perform superior on the nights with better seeing. It will happen that on one night with excellent seeing, you'll be able to find a particular deep-sky object, while on the next night, with poor seeing, the object remains invisible. You have not become a poor observer, nor is your optical equipment malfunctioning. It's just the seeing that is affecting your observation. That's why you should take note of your local seeing. You can express the seeing in a scale of five grades: excellent – very good – good – average – poor.

Usually transparency and seeing don't go well together. The nights with best seeing, when the atmosphere is very calm, are often of high humidity and thus less transparent. The windy nights, when the air is very clear, can often offer a higher nelm but at the cost of better seeing. And last but not least, try to observe the desired deep-sky objects when they are at their highest in the sky. That is where you will find the thinnest layers of air.

Your Optical Equipment

Telescope or Binoculars?

It's a common believe that a self-respecting astronomer needs a substantially large telescope if he or she wants to make any kind of decent observation. The truth is that there is no such thing as a perfect telescope, one that suits every need. Some telescopes offer a wide field of view (the richest field telescopes), while others are capable of delivering very high magnifications.

If you are a novice observer, you're better off with a wide field telescope. How about binoculars then? Well, they are excellent wide-field devices. This is good news for the readers who already have a common pair of binoculars, and even if you don't have one of your own, maybe you know someone from whom you can borrow one. Do you need a telescope then? Well, for your journey through the universe, a pair of 7×50 or 8×40 binoculars will do nicely. Binoculars are great devices for novice observers for several reasons. They require no set-up time, nor cool-down time. You just take off the lens covers and off you go. They display an upright view, compared to the confusing mirrored or upside down telescopic views. And they offer the widest possible view, which is very convenient for finding objects. Add to that the fact that binoculars are lightweight and very portable, and you have the best mix of telescopic qualities that are mandatory for a novice observer.

Binoculars are in fact two refractive telescopes mounted together. They were invented right after the telescope. The fact that we can use our two eyes for observing becomes more important at night. Our night vision is distracted by "noise" when we observe faint objects. The human brain is very capable of "merging" the images each eye generates into a single image with less noise. All this may sound very technical, but just do the test and you'll see what we mean: The next time you go observing at night, close one eye and count the stars near the zenith. Next, open both eyes and count again. The difference is amazing. That's why we see better with binoculars than with a monocular of the same size.

Binoculars come in a variety of designs, sizes and magnifications. Binocular designs have greatly improved over the years, to get to the two popular

Fig. 1.15 These are small porro-prism binoculars. The eyepieces have rather large lenses, which are very convenient for bespectacled observers

types of today: the porro-prism models and the roof-prism models. The roofs have the more compact and lighter design but are more expensive. Both types make use of prisms to fold the light path and to obtain an upright image. Without these prisms, the binocular's tubes would be much longer. The choice between roofs or porros is up to you, as both types can deliver the same image quality (Fig. 1.15).

Binocular manufacturers have a distinctive way of displaying their specifications: magnification × aperture (see Fig. 1.17). 8 × 40 binoculars magnify eight times and have front lenses (aperture) of 40 mm diameter. Thus 7 × 50 binoculars magnify seven times and have an aperture of 50 mm. All kinds of models are available: 8 × 20, 7 × 35, 8 × 56, 10 × 50, 10 × 70, 15 × 70, etc. The higher magnification and greater aperture sound attractive. But be aware that aperture comes with weight (and price!) and that higher magnification is hard to hold still. If you want a larger pair of binoculars than a 10 × 50 you'll need a tripod to mount them on.

The most important function of an astronomical telescope is to gather light. The night-adapted human eye has a dilated pupil diameter of a maximum of 7 mm. With age, the pupils dilate even less. In other words, the maximum aperture of our eyes is 7 mm. A pair of 8 × 40 binoculars have an aperture of 40 mm. Therefore, 8 × 40 binoculars gather 33 times more light than the naked eye. When you put these binoculars in front of your eyes, you simply have widened your pupils from 7 to 40 mm! That's the main reason why you see so many more stars with a pair of binoculars. So more aperture is better, right? How about an 8 × 150 binocular then?

Well, aperture comes with a cost. Although it's perfectly possible to make such an 8 × 150 instrument, it would not perform any better than an 8 × 56 binocular! The reason is that every telescope generates an exit pupil. The exit pupil is an optical plane behind the eyepieces where all the gathered light of the telescope is composed into an image that can be examined by our eye. The larger the exit pupil, the brighter the image. The diameter of the exit pupil equals the aperture divided by the magnification.

The 8 × 40 generates an exit pupil of 40 divided by 8, which equals 5 mm. The 7 × 50 has an exit pupil of 50 divided by 7 equals 7.1 mm. An 8 × 150 would generate an exit pupil of 150 divided by 8 = 18.75 mm.

We've just learned that the human pupil dilates to about 7 mm, so a lot of the gathered light of the 8 × 150 would be blocked by the iris of our eye. So it makes no sense to generate an exit pupil that is larger than the pupil of the observer. The manufacturer of the 150 mm aperture binoculars will have to deliver a magnification of minimum 150 mm/7 mm = 21.7. Otherwise, the customer would not be able to use the full aperture of his or her binoculars. Therefore, the bigger the aperture of our instrument, the higher the magnification should be, to use the aperture to its full potential.

Now, let's go back to our 8 × 40 and 7 × 50 binoculars. If your pupils widen to 7.1 mm, you would be able to use the full 50 mm aperture of the 7 × 50 binocular, but also those of the 8 × 40 and the 10 × 50 binoculars. A smaller exit pupil delivers a dimmer image; nevertheless we still make use of the

Fig. 1.16 Binoculars come in all kind of sizes. These porro-prism models are 8×40, 10×50 and 8×56

full aperture of our binoculars. However when the view of the sky is hampered by light pollution, or your eyes are lit by a nearby streetlight, your pupils only dilate to maybe 5 mm. In that case, your pupils would not be able to collect all the light that is gathered by the 50 mm aperture of a 7×50. You could better use a lighter pair of 7×35 binoculars then, or choose between the 8×40 and 10×50. This also explains why for daytime use, when our pupils remain small, the smaller aperture of let's say 8×20 binoculars is perfectly acceptable (Fig. 1.16).

Besides aperture and magnification, there is one other parameter that manufacturers specify in relation to their binoculars – the true size of the fov. The true fov can be specified in degrees (see Fig. 1.17), in meters, or in feet or yards. A pair of 8×56 binoculars have a true fov of nearly 6° (see Fig. 1.12). From one end of the field stop (the black edge in the eyepiece that narrows the field of view) to the opposite end, the amount of sky that is displayed measures 6°, although that amount of sky is magnified by the

Fig. 1.17 The specifications of the binoculars are usually printed on the prism housings

eyepieces by 8 to an apparent fov of 6°×8, which equals 48°. It means that these binoculars show a field stop that measures 48°, in which a true field of view of 6° is presented. When you look at the Moon (which measures 30′) with these binoculars, you see in the eyepieces a Moon disk that measures 30′ × 8′, which equals 240′ or 4°.

All eyepiece designs have image flaws near the edges of the fov. Field stops are necessary to hide the blurry edges of the generated fov. The more expensive binoculars are equipped with higher quality eyepieces (and sometimes bigger prisms) to deliver a sharper fov. They are capable of displaying a larger fov that is still sharp near the field stop. Beware of manufacturers who offer cheap binoculars with a relative large fov. The outer fov might turn out to be useless in the field.

If you intend to buy a pair of binoculars, if possible, try them out before you make a choice. Especially those who wear glasses should try before they

Fig. 1.18 Binoculars with multicoated lenses transmit more light and provide brighter images

buy. You might want to keep your glasses on when you observe. Some binoculars are better suited for that purpose than others. The eye relief specifies the maximum distance between the eyepiece and the eyeball of the observer at which the whole fov is still visible. If you hold the binoculars at a longer distance than the eye relief, you'll see a narrower fov. Look for binoculars with eyepieces with a long eye relief of 17 mm or better.

Also take a good look at the coatings of the lenses. Coatings are essential to minimize light reflections in the optic path. Look for the specification "fully multicoated." Uncoated lenses transmit less light than fully multicoated ones. Just hold the binoculars at arm's length with the front lenses turned to you. If you can clearly see the reflection of your own face in the front lenses, then the coatings are not good. Now turn them around and do the same test with the eyepieces. Coatings should look dark, with greenish or bluish tints, without showing any flaws or spots (see Fig. 1.18). If possible, try several samples and pick the one that looks the best. Ask the vendor to check and to adjust the collimation of the prisms. Also check the focusing

Fig. 1.19 Check the eyepieces as well. They, too, should be multicoated. Adjustable eyecups are very convenient for observers with glasses. Large eyepiece lenses are preferable

mechanisms. If they are too stiff or too loose, they'll turn out to be useless in the dark (Fig. 1.19).

Binoculars can be bought everywhere, but it is best to buy in a specialized (online) shop where every sample is checked and adjusted by experienced people. Maybe they'll cost a little more, but you'll know that you get the quality that you pay for.

All the deep-sky objects in this book are visible with pair of modest 8×40 binoculars and many of them are just better in a pair of binoculars than in a telescope. Now don't think that we're against using telescopes – on the contrary! Telescopes are wonderful instruments once you know how to use them. However, they are heavier and more complex to set up and to use, because of the specialized mounts they require. Some types of telescopes (the reflectors) need to be optically aligned before you can use them. The larger telescopes are equipped with a finder scope, which needs to be aligned as well. And telescopes need to adapt to the outside temperature before they show any decent images. All these requirements can be very overwhelming for a novice observer who does not even know his or her way around in the night sky. Many new telescopes gather only dust instead of light because their owner never really got into stargazing. If you only have one evening hour free to do stargazing during the working week, don't even think of setting up a telescope. You're better off with a pair of binoculars for starters. Once you get the hang of it, you can always purchase a fine telescope later.

If you already have a telescope, you can use it as well, although with the lowest possible magnification.

Additional Equipment

Although a pair of 7×50 and 8×40 binoculars are relatively effortless to hold still, you might prefer to mount them on a tripod (see Fig. 1.20). The tripod allows you to examine deep-sky objects for a prolonged time and leaves your hands free to take notes or to make a sketch. Most binoculars can be attached to an ordinary photo tripod using an L-adapter.

You might also like to use an observing chair. Comfort is important when it comes to observing faint fuzzy objects high in the sky. If you sit down you'll be more focused on observing than on keeping the binoculars still. A lawn chair (especially the type with armrests) can be very convenient to observe objects near the zenith.

You should also invest in a small red LED flashlight to examine the finder charts in this book. The red LED will preserve your night vision. You can also use red tissue paper in front of a regular flashlight. Don't use too much light, though, because it will ruin your night vision.

A small foldable table will be very convenient, too. If you observe from your own backyard, you probably already have a table and a chair outside, waiting for you.

Don't forget the bug spray for balmy summer nights as well. If you're planning to stay up for a couple of hours, you could bring along some candy bars, together with some hot or cold drinks, depending on the season. Avoid alcohol and tobacco; they don't mix up very well with keen night vision.

Fig. 1.20 Binoculars can be attached to a tripod with the use of an L-adapter

Dress for the Night

One of the greatest enemies of the amateur observer is the cold. Temperature drops quickly once the Sun has set, especially on the clearest of nights that are ideal for stargazing. When we're not properly dressed, our bodies lose heat very quickly. This hobby of ours involves little physical activity to keep ourselves warm. And once our hands or feet get cold, there's no way back, unless back inside. Even summer nights, you should wear warm shoes, denim jeans and a hooded sweatshirt plus a light jacket. During autumn and spring, you can add the old-fashioned but very effective long underwear, snow boots, a warm wool jacket with hood and a pair of warm wool gloves. Winter viewing can be really challenging, especially when it's freezing cold. You might want to add a lined ski suit and an extra wool cap. It also helps to drink a cup of hot chocolate or hot soup just before the observation session. When you observe from a remote site, bring along an insulated can with hot coffee or tea to keep warm during the night.

Vision Under Low Light Conditions

The human eye is a wonderful instrument that is capable of functioning under a wide range of lighting conditions, from scorching bright daylight to the dark of the night. If you know how the eye works, you can take advantage of it when you're observing at night.

The most important part of the eye is the retina, a filmy piece of tissue inside the eyeball. The retina contains two types of photoreceptors, the rods and the cones. The cones are the receptors that are responsible for the sharp, bright and colored daytime vision, while the rods are more sensitive receptors that are used for low light vision.

The cones and rods are not evenly distributed over the retina. The cones are highly concentrated in a spot right behind the eye lens, the fovea. The fovea is responsible for our sharp daytime vision. When we examine our environment, we instinctively move our eyes in a manner that projects the object of our interest onto the fovea, where our vision is the sharpest. This is called direct vision. You may not be aware of it, but right now you are projecting these sentences onto your fovea. Did you notice that when you focus on the first line of this page, you cannot read the bottom line of the page? You can see the whole page, but only the part you're focusing on is readable. This is because the cone density becomes very low outside the fovea. Our vision is not consistently sharp over the whole field of view. The fovea is so densely packed with cones that it is void of rods. The rods are distributed around the fovea. They provide us the peripheral black and white vision under low light conditions. The retina also contains a blind spot, where no receptors are located. This is the place where the optic nerve leaves the eye. The blind spot should be avoided while stargazing. Objects projected at the blind spot remain invisible.

Our night vision is very dependent on the information gathered by the sensitive rods. Unfortunately, novice observers have no experience with viewing at night. We have just mentioned that the fovea is devoid of rods, and that therefore, when we focus on a dim object, we will not see it. Our direct vision fails.

The cones are not sensitive enough to be triggered by low light levels. This means that the whole fovea is practically a blind spot, too. The right technique to use here is called "averted vision." We have to project the object of our interest onto the rod-rich part of the retina, which is located around the fovea (see Fig. 1.21). It appears counterproductive at first, but you better let your gaze wander around the object than to look straight at it. It takes practice to look at one direction while paying attention to another. That's why experienced observers see more objects and more detail in objects than novices do.

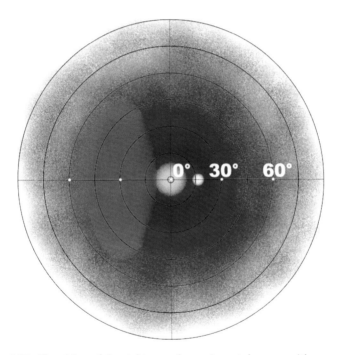

Fig. 1.21 The vision of the right eye when using a telescope with one eyepiece. The cones are concentrated in the yellow area, called the fovea. This is the area we tend to use when we look at objects. The rods are distributed around the fovea, in the blue area. Dim objects are best viewed in the *dark blue* ring around the center of view, where the concentration of rods is the highest. The white spot, located at 18° right of the center of view, is the blind spot. Objects should not be viewed at that region. The *dark gray* oval represents the place where a right eye user should place dim objects for best averted vision. In other words, the observer should look slightly right of the object. Left eye observers can mirror the diagram

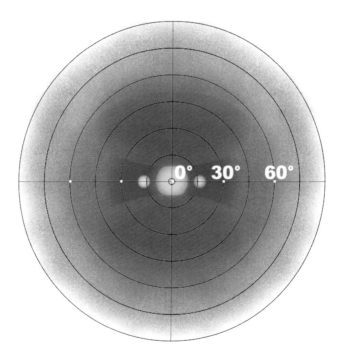

Fig. 1.22 The merged vision obtained from both the left and right retina when using binoculars or a telescope with a binoviewer. The view contains the blind spots from each retina. The *dark ovals* represent the area where objects should be placed for best averted vision. Binocular observers should try to look slightly above or beneath the object of interest

The more time you spend behind the eyepiece, the better you'll see. Observing is an acquired skill. Seasoned observers do place the object of their interest in the so-called "sweet spot." The sweet spot is the most sensitive area on your retina, where you find detecting dimmer objects or faint stars easiest. The location of the sweet spot is different for every observer. If you observe with a telescope with one eyepiece, you can use Fig. 1.21 to avoid the blind spot. The dark gray oval is a good starting point to seek your own sweet spot.

Binoculars allow us to observe with both eyes. Our brain perfectly merges the visual information of both retinas into a single improved image. That's why binocular vision delivers brighter and sharper images. Binocular users should take into account that their field of view contains three spots where dim objects are obscured (see Fig. 1.22).

The human retina is more sensitive for a moving stimulus. If you keep your eyes focused at a fixed point, your vision will turn gray. Try to keep your gaze wandering around in the fov of the eyepiece. The deep-sky objects will become easier to see. And don't forget to keep breathing while observing. Some people hold their breath, and their oxygen-starved retinas will return a gray image.

Preserving Our Night-Vision

The rods contain large quantities of a molecule called rhodopsin. When a rod is exposed to light, its rhodopsin bleaches, whereby the rod transmits a neural signal to the retina. If the rods receive too much light, like under daylight conditions, all their rhodopsin is bleached and the rods become inactive. Only when it's sufficiently dark do the rhodopsin quantities build up again, and the rods become sensitive. In practice, we notice this when we come in to a dark area from a well lit room. At first we can't see anything. The pupils will dilate, allowing the cones to receive more light. This happens within a few seconds. The cones will become a little more sensitive as well after a few minutes.

But our eyes are not really fully dark adapted. The process of building up rhodopsin is a slow one. Only after 30 min under low light conditions are the rods sensitive enough to function. And after 2 h they have reached their maximum sensitivity. It is important to protect the slowly gained sensitivity of your rods. Once your eyes are fully dark adapted, you should avoid using any bright light, or looking into a bright light in the neighborhood. But when you want to consult the finder charts, or if you want to take notes, you need your cones to function, too, without bleaching too much of the rhodopsin out of your rods. The solution is a red light. Rhodopsin is very sensitive to green light, but less sensitive to red light. In the dark, greenish objects look brighter. If we use a small red penlight, our direct vision will allow us to read and write, but the red light will not bleach large quantities of the vulnerable rhodopsin. Remember that going back into a well lit place will bleach all the rhodopsin. The whole process of dark-adaption will have to repeat itself. Even the dashboard lighting of your car as well as the display of your cell phone is too bright. Some people like to use their laptop in the field. There are special red screens to dim monitor screens, too (Fig. 1.23).

Fig. 1.23 The light domes created by villages and cities can ruin your night vision. Avoid sites where you can directly look into street lights and the like. This picture is made at an observing site looking south in the direction of Scorpius. Note the delicate structure of the Milky Way left of Scorpios. A few meters away, the trees block the lights and vision can fully dark adapt

Direct and Averted Vision

While we observe, we constantly utilize a mix of direct and averted vision. Our direct vision allows us to see bright objects with a higher resolution, like close double stars. Our more susceptible averted vision allows us to observe faint objects, though with much less resolution. Some objects are suitable for both direct and averted vision. Many star clusters have brighter members, which are visible with direct vision. Their dimmer counterparts can usually be discerned with averted vision. And on many occasions, the clusters remain unresolved. Subsequently we can only see the combined glow of the individual cluster stars with averted vision. Nebulae and galaxies also benefit from the combined use of direct and averted vision. Direct vision will show the brightest features in high detail. Averted vision will reveal the dimmest extensions of these objects. That's why our visual impression of the deep-sky from behind the eyepiece is completely different from the misleading flashy full color images taken with an astronomical camera. Therefore you will not find lots of photos for describing deep-sky objects in this book. All the objects are accompanied with a sketch that is based on a real visual observation.

The Optical Performance of Binoculars and Telescopes

The naked eye can only see a fraction of all the stars in the sky. This limit is called the nelm. With our binoculars and telescopes, we provide our eyes with larger pupils. The stars appear brighter and our eyes succeed in detecting stars beyond the nelm. But how faint a star can we see? It is possible to predict the telescopic limiting magnitude as well. It's important that an observer knows what he or she can and cannot expect to see with optical equipment. Table 1.4 shows what the telescopic limit is according to the nelm, the aperture and the magnification of your telescope.

If you take a close look at Table 1.4 you see that with growing aperture, the limiting magnitude grows, too. The table also shows that a small 7×35 pair of binoculars at a dark site (nelm 6.0) performs as good as a heavy 15×70 pair in a light polluted area. The 15×70 perform even worse if your eyes

Table 1.4 The calculated limiting magnitude (lm) of a telescope in relation to the nelm

	nelm	4.00	4.50	5.00	5.50	6.00
Telescope	7×35	7.47	7.97	8.47	8.97	9.47
	8×40	7.76	8.26	8.76	9.26	9.76
	7×50	7.85	8.35	8.85	9.35	9.85
	10×50	8.24	8.74	9.24	9.74	10.24
	15×70	9.50	9.55	10.05	10.55	11.05
	100×25	9.99	10.49	10.99	11.49	11.99

aren't fully dark adapted. It also demonstrates that if you want to see a mag 8.7 object with an 8×40 you need a site with a nelm equal to or greater than five. The table also explains why some clusters resolve and others do not. If you are observing an open cluster with an integrated magnitude greater than the limiting magnitude (lm) of your scope, the combined glow of the cluster members will be visible in the fov as a smudge of light. Resolution takes place as soon as individual cluster members shine brighter than the lm. In that case you'll be able to discern discrete stars within the cluster's background glow. Stars at the verge of resolution can give a cluster a mottled appearance. Because the nelm and the lm can change from night to night, so can the observed appearance and resolution of a cluster change.

The majority of the deep-sky objects described in this book are brighter than mag 8.5, with the exception of a few galaxies. You can use Table 1.4 to check what objects are within the reach of your lm.

It takes time and practice to determine the nelm. You can use the seasonal sky maps for a rough estimate. These sky maps show all the stars up to mag 4.5. So if you can see all these stars in the sky, your nelm is equal to or better than 4.5. If can see the four corners of the bowl of the Little Dipper (Fig. 1.11) then your nelm is at least five.

Observing at Last!

If you want to put all this information into practice, the following checklist could help you along.

The perfect observation session starts with a decent preparation before you leave the house.

- Check the weather reports and the phase of the Moon.

- Check (online) the position of interesting planets and comets, and passes of bright satellites.

- Choose the appropriate observing site.

- Check the binoculars or telescope.

- Check the red penlight (batteries).

- Check the mount, binocular adapter, table and observing chair.

- Check the blanket that you can lay over the hood of your car. It makes a perfect table.

- Check that you have your finder charts and this book.

- Make a list of the objects you want to see.

- Check your notebook (logbook)and pencil, and maybe a sketch board.

- Adjust your watch.

- Check your clothing (gloves, hat) for the night, plus that you have the bug spray.

- Check that you have the food and the (warm) drinks.

- Inform someone on your whereabouts.

- Check your cell phone.

- When observing from the backyard, put out the lights and close the curtains. Ask your fellow inhabitants to refrain from producing any light in the rooms that have windows to the backyard.

- When observing from a remote side, check how to dim the light in the trunk of your car, or put a piece of red cellophane over it.

The list is quite long, and you don't need to do all of it when you're just going out in your own backyard. But even if you just observe from your backyard, your night vision might get ruined if you need to get something from inside.

When you start your observing session, the first thing to do is get your bearings right. From the Big Dipper, find the Pole Star. Then look up the appropriate seasonal map. It's a good practice to compare the constellations on the seasonal maps with the real ones. This is time well spent while your night vision builds up. Next, estimate the seeing and the transparency of your sky and make a note of it in your logbook. A small logbook is enough. It is not feasible to remember all the deep-sky objects you saw. The visual impression might look very strong, but after a day or two, the visual memory will fade away. The logbook will allow you to compare your past observations with the new ones. Besides, making descriptions puts your visual acuity on a higher level. Some observers prefer to use a small voice recorder instead. They write down notes at a later time.

Once you've located the constellation of interest, you can study the finder chart of the desired deep-sky object. Compare the stars in the chart with the ones in the sky. The finder charts will show fainter stars than the nelm. This is done on purpose, because the lm of your 'scope is better than the nelm. Now that you've found where to look, take out the binos or the telescope. Try to locate the brighter stars of the finder charts in the eyepiece, and find your way to the deep-sky object.

Now that you've found your target, congratulate yourself with the positive hunt and then – hold your horses. Deep-sky observing is not a race. Try to spend as much time as possible with your deep-sky object and let your gaze wander around in the eyepiece. Because of the low light level, your brain and eyes need a lot more time to interpret the view in the eyepieces. Your first glance at a deep-sky object might disappoint you. Remember that seasoned observers see more than newbies. First try the direct vision, and then switch to averted vision. Try to look at the deep-sky object for at least 5 min. Then start to take notes.

Your notes should be comprehensive, as if written for someone who never saw the object. This really sounds silly, but you risk relying too much on your visual memory. The memory will fade away but not the notes. The longer you look, the more the object will grow in size and in detail. Why not make a quick sketch? Sometimes, a sketch says more than a page full of notes. When you sketch, note the orientation of the view: where is north (towards Polaris) and where is west (the direction of the drift of the stars). The sketching will improve your vision because you are forced to pay attention to small details that were overlooked at first. After all, a sketched object will leave a stronger impression because you made a more profound study of it.

Sometimes a dim star or a faint feature disappears. It could have been your imagination that's playing tricks with you. Maybe the seeing is troubling your view. Just wait and see if the star or feature pops back into view at the same location. If you can see it for a third or fourth time, then it's really there.

Once you're convinced that you had a really good look, start over with a new object. It can also happen that you cannot find the object at all. Do take note of this negative observation, for future purposes. You can always try the object on another night.

Objects are more interesting when they are high in the sky. Don't expect to see much detail in objects that are lower than 15° above your horizon. You can organize your list of objects by starting with the ones in the west. They will set first. Then move through your list eastwards.

It's worthwhile to look again at objects you've already seen before. The second look often is more pleasing and richer in detail. The eye sees better in a familiar environment. Soon you'll have your own favorite seasonal highlights, the objects you like to visit over and over again.

Join the Club

While you're out there under the stars, you are not alone. Each clear night, thousands of amateurs are watching the stars. Perhaps some of them are not too far away? Why not observe together? Maybe you can join a local astronomy club. Your best starting point is the Internet. There are several national and international web-based forums where amateurs meet. Observing in a group is very appealing. You're not alone in the dark. You share experiences, views and even equipment with each other. You'll also discover new objects and better observing sites. See the section on Internet links in Chapter 15.

About the Sketches

The sketches in this book represent the author's personal visual impressions of deep-sky objects with various binoculars and small telescopes. Sketches are a better way than photographs for showing in a very subtle way what deep-sky objects look like in the eyepiece of the observer. When you just take a quick look at the sketches, not all the features will jump directly to the eye. You'll have to study the sketches if you want to see all the details, just like in a real eyepiece.

The sketches and the finder charts in this book all have an upright orientation–north is up and west is to the right. This orientation will not always match with the orientation of your observing site. When you observe with binoculars, rotate the top of the book towards Polaris. Observers with a telescope can find north in the eyepiece by moving the tube of the 'scope towards Polaris. The side where new stars enter the field of view is north. The direction towards the stars' drift (when the 'scope is not motorized) is west. If your telescope is equipped with an erecting prism, the view will be mirrored compared to the sketches.

Sketching is a very useful way to study deep-sky objects. They are perfect tools for sharing your experiences with others. The drawing forces you to keep your attention on a high level. When you sketch, you see more, and you'll become a better observer (Fig. 1.24).

Now don't expect sketching to be easy, especially when you're not used to dark-adapted observing. Once you feel comfortable behind the eyepieces, and the object is not too complex, you can give it a try. Starters can use white paper and a graphite pencil. Some experienced observers use black paper and white pastel pencils. Included is a picture of the sketching template that the author uses (Fig. 1.26). The drawing circle represents the field of view. Remember that you cannot use too much light. Therefore a large format is easier to read at night. A large circle may look intimidating at first, but it gives you plenty of space to add the details in an accurate

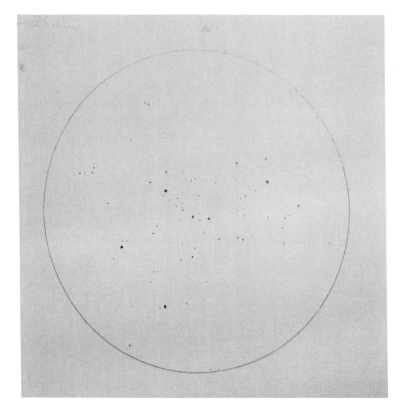

Fig. 1.24 A pencil sketch of the Butterfly Cluster, M6. The sketch is based on an observation made with a small telescope

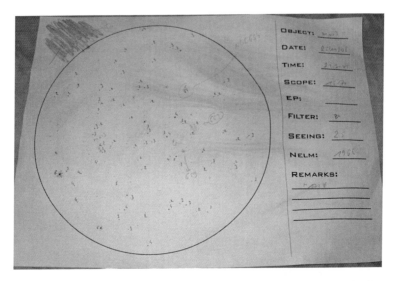

Fig. 1.25 The field sketch when finished. The circle that represents the fov might look excessively large. In practice it's easier to draw on a larger scale in the dark. Note the numbers added next to the stars. They represent the comparative brightnesses of the stars. This information is used to complete the digital sketch

way. Start first with the brightest stars. These are the anchors of the sketch. For them, you can use your direct vision and a little more light on the drawing board. Stars are drawn as black dots. The brighter the star, the larger the dot. The sketch becomes a negative image of the fov. Bright features are represented as dark gray on the paper. Once these stars are positioned, use them as a reference for pointing the dimmer stars. The dimmest features come at the very end. It's is not mandatory to draw all the stars in the field. You can add plenty of notes on your sketches later. Because of the low light level in which you draw the sketch, you may not be able to transfer many of the subtleties you see in the eyepiece onto your drawing paper (Fig. 1.25).

These subtle features are more fully described in the supplementary notes. When the observing session is over, make a clean digital sketch on your computer, based on the information of the field sketch. Digital sketching

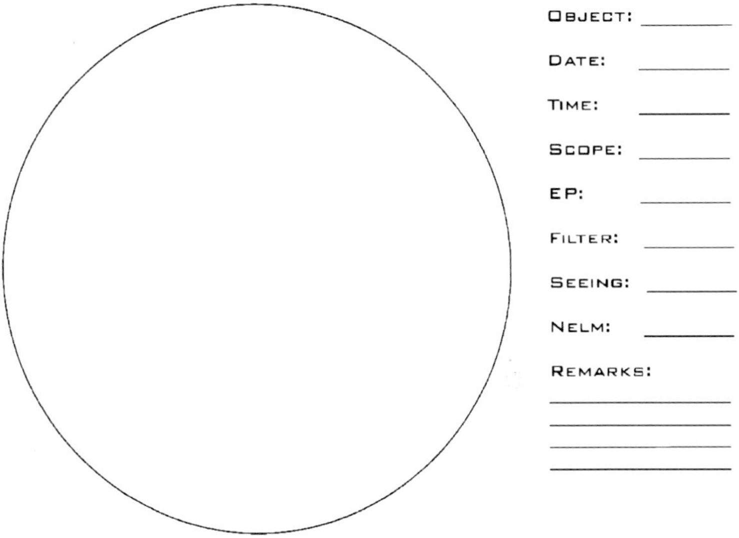

OBJECT: _____

DATE: _____

TIME: _____

SCOPE: _____

EP: _____

FILTER: _____

SEEING: _____

NELM: _____

REMARKS:
_____
_____
_____
_____

Fig. 1.26 This is the template the author uses for sketching deep-sky objects

becomes very easy when a sketching tablet is used. You can take a picture of the field sketch and use it as a background in the drawing program. With the sketching tablet, you can draw all the features in a clean way on multiple layers over the background. When all is drawn, you then delete the background. Finally you invert the sketch. The black dots on a white background become stars in the sky.

CHAPTER 2

WELCOME TO THE MILKY WAY

Fig. 2.1 The Sun is a small star located near the rim of one of the spiral arms of the Milky Way. It's located 30,000 ly from the center of our galaxy

Introduction

Observing the stars is a great pastime. The pleasure becomes even greater when we comprehend our vantage point and what the objects we are looking at represent in terms of parts of a greater structure, like the Milky Way. Most classic astronomy books project the universe onto a celestial sphere, a huge hemispheric dome, with the observer at the center of it. The concept of the celestial sphere is very useful to explain the motions of

R. De Laet, *The Casual Sky Observer's Guide*, Astronomer's Pocket Field Guide, DOI 10.1007/978-1-4614-0595-5_2, © Springer Science+Business Media, LLC 2012

the heavenly bodies such as the Sun, the stars and the planets. We deliberately did not follow this approach in this book because imagining all the celestial objects on a dome takes away the depth of field.

You will soon understand the celestial motions through your observations in the field. With the use of various figures, sketches and galaxy views in this book you should gain an extra sense of depth of field when you are out observing the stars. Observing is obviously done with your eyes, but even more it is done with your mind's eye. On our voyage through the universe, the various galaxy views explain the true location of the deep-sky objects in space. Once you understand the true depth of space, you will be able to see various structures within the universe and the Milky Way in real perspective. But first we need to know our place in space.

Let's start with our Sun. Not so long ago, the Sun was believed to be the center of the universe, but it is not. The Sun lies at 30,000 ly from the center of the Milky Way. Our day star is a modest star, one member of our home galaxy. Today astronomers catalog our Sun as a small G2V star with an absolute magnitude of 4.8 and an estimated fuel (hydrogen) reserve of 4–5 billion years. The Sun orbits the center of the Milky Way, just like Earth orbits the Sun. In fact all the stars are in orbit, otherwise the Milky Way would collapse under its own gravity. One orbit of the Sun around the center of the Milky Way, which we can call one galactic year, takes about 220 million years. Given its age of 4.5 billion years, it has made 20 or so revolutions (and Earth with it), which equals an age of 20 galactic years. The Sun will continue to do so for another 20 galactic years, before it runs out of fuel.

Until recently, our species was tied completely to Earth. We've only made it to the Moon and back. Other destinations seem inviting. Mars is a worthy candidate for our next stop in space. Nevertheless manned space exploration turns out to be a complicated endeavor. And yet, we can consider ourselves interstellar travelers. Our trusty spacecraft is called Earth, propelled through the universe along with our Sun. It's not a frivolous idea, really. Since the dawn of modern human culture, about 50,000 years ago, the Sun, and our kind with it, has traveled 42 light-years through space. It's an exhilarating leap in interstellar space covered by humankind in the safest possible spacecraft ever built. About 42 light-years is the distance to Capella, the brightest star in the constellation of Auriga. We may have had close encounters with other stars, considering that more than 300 stars are closer to us than Capella.

All the stars we see in the sky with the naked eye are distant suns, belonging to the Milky Way. Astronomers estimate the total number of stars in the Milky Way at over 300 billion. Sizes and numbers are overwhelming in space, but a galaxy like the Milky Way is still a very empty place. When two galaxies merge, there is so much clear space that their stars never collide.

We can describe the Milky Way as a huge rotating disk of smoothly distributed (mainly A, F and G) stars. The disk is about 3,000 ly thick and measures 100,000 ly in diameter. At the center of the disk, there is a small but dense central bulge, consisting of older, evolved (K and M) stars.

The center of our galaxy is relatively poor in gas and dust. The disk of our galaxy contains the spiral arms, which are very rich in gas and dust. These spiral arms are about 1,500 ly thick. They also contain the emission nebulae and the young and hot (O and B) giant and supergiant stars that light up the spiral structure (see Fig. 2.1). The Milky Way is surrounded by a halo of globular clusters. These globulars orbit the center of our galaxy on elliptical paths, whereas the disk stars have a more circular orbit within the disk. This implies that the globular clusters must pass through the disk to complete their orbits. The disk stars also have a different composition than the globular stars. Astronomers call disk stars, like our Sun, Population I stars. Globulars are generally made up of Population II stars. Population I stars are said to have a higher metallicity than Population II stars. In astronomy, all elements heavier than helium are called metals. The metallicity is a measure of the age of a star.

In the early evolution of the Milky Way, the interstellar medium was thought to be low on metals. Stars formed at that time would have a low metallicity. When the Milky Way grew older, the metallicity of the interstellar matter increased due to the heavier elements that enriched the interstellar medium by means of planetary nebulae and supernova explosions. The later a star forms, the higher its metallicity is.

Because of our vantage point from within this huge disk, we see the Milky Way like a glowing path of unresolved starlight, like the backbone of a huge creature that bends over our heads. The stars that make up the classical constellations are merely disk stars that are closer to us, whose light is bright enough for us to see them resolved against the background glow of the sky. Unfortunately our view of our home galaxy is greatly obscured by huge clouds of dark interstellar matter of which the spiral arms are made up. To us, the Milky Way resembles the look of an edge-on galaxy, whose central bulge is also obscured by its spiral arms. Without this obscuring interstellar matter, the Milky Way would appear much brighter. If a supernova explosion would occur near the center of our Milky Way, we could not see it, because all of its light would be absorbed by the interstellar dust. Ironically we can in fact examine nearby galaxies much better than our home galaxy. However there exist several windows in the spiral arms, places where the interstellar clouds are much thinner, which we can and will use to look deeper into our own galaxy.

Our knowledge of the true structure of the Milky Way remained limited for a long time because the large interstellar dust clouds are opaque to visible light. With the development of radio telescopes and infrared telescopes in the 1950s, astronomers could see through the interstellar medium. The appearance of our galaxy, as in Fig. 2.1, is a calculated guess, based on the latest observations at various radio and infrared wavelengths.

The Galactic Coordinate System

We can locate the structures and the deep-sky objects within our Milky Way with the use of a coordinate system, just like the coordinate system we use on Earth. Astronomers make use of the Galactic Coordinate system ('galactic' refers to our Milky Way). We can imagine a galactic plane in the middle of the galactic disk. The center of the central bulge of our galaxy also lies on this galactic plane. The projection of this galactic plane in the sky then represents a line that we can call the galactic equator.

From our point of view, within this galactic plane, the galactic equator resembles a grand circle projected in the sky, with us in the center of it. The galactic equator traces the middle line of the Milky Way across the sky, as represented in Fig. 2.7. On the galactic equator we can measure the galactic longitude from 0° to 360°. We agree that 0° longitude points to the center of the Milky Way, in the constellation of Sagittarius. The anti-center, at 180° longitude, points to the edge of the Milky Way on the opposite side of where we see the center, in the constellation of Auriga. 90° longitude

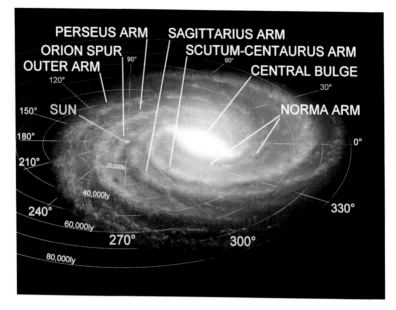

Fig. 2.2 The galactic plane, with its longitude directions and distance scale, lies in the middle of the galactic disk. The various spiral features bear the name of the constellation in which they are prominently visible

lies in the direction of the constellation of Cygnus. This is the direction at which the Sun is orbiting the hub of our galaxy. Opposite of 90° is 270°, the point in the constellation of Vela, where the Sun came from, on its orbit around the center of our galaxy. Figure 2.2 displays the galactic plane and the longitude measures astronomers use. All the galaxy views throughout the book have the same orientation as Fig. 2.2.

Galactic latitude indicates the angular 'height' of an object measured from the galactic equator. It ranges from −90° to 0° to 90°. There are two important directions: the galactic poles. The galactic North pole (at 90° latitude) lies in the direction of Coma Berenices. The galactic South pole lies in the direction of Sculptor. When we look in the direction of Coma Berenices, we look out of the galactic disk where our view is less obscured by the interstellar medium.

Our Orientation in the Milky Way

The galactic coordinate system is very straightforward. However, the orientation of Earth in space is by no means related to the galactic coordinate system. But what is then the orientation of Earth, with respect to the galactic plane? Figure 2.3 tries to illustrate the Earth-based view of the Milky Way.

The orientation of Earth is linked with the position of Polaris. Polaris is a disk star with the galactic coordinates 123° longitude, 27° latitude. For convenience let's suppose that the rotational axis of Earth is pointed towards Polaris. The 123° longitude is in the direction of Cassiopeia. You can check also the seasonal map for spring (Fig. 1.4) where Cassiopeia hovers above the northern horizon. During spring evenings we have Coma Berenices and thus the north galactic pole near the zenith. When we look in the direction of Cassiopeia, we look in the direction of the galactic plane around 123° longitude.

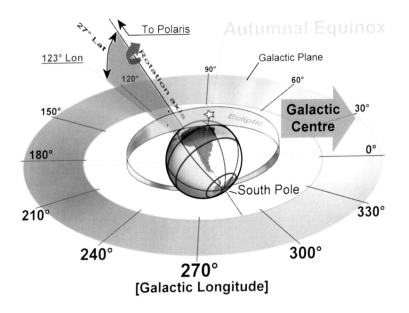

Fig. 2.3 The orientation of Earth with regards to the galactic plane. The *golden ring* around Earth represents the ecliptic, the apparent path of the Sun in the sky. Both the galactic equator and the ecliptic are *grand circles* in the sky

When we look at Coma Berenices, we look in the direction perpendicular to the galactic disk, thus at the north galactic pole. The golden ring is added to Fig. 2.3 to represent the path of the Sun among the stars – the ecliptic.

With the change of the seasons, the Sun's position and thus our nighttime window to the universe also changes. During winter, when the Sun is on the right side of the golden ring (near 0° longitude), it appears low in the south for a mid-northern observer. This indicates that the opposite site, near 180° longitude, is visible during the night. That's why we can see the part of the Milky Way in the direction of Auriga during winter (see also Figs. 1.9 and 1.10). In spring, the Sun is located in the part of the golden ring that is obscured by Earth (see Fig. 2.3), beneath the galactic plane. Therefore the nights of spring offer us a clear view at the north galactic pole, and less clear of the galactic plane. When summer comes, the Sun is near 180° longitude, at the left side of the golden ring. Then the galactic plane from 90° longitude to 0° longitude is visible at night (see Figs. 1.5 and 1.6). And finally, during autumn, the Sun is located near the star symbol on the golden ring. During that season, the Milky Way is visible at night from 30° longitude up to 160° longitude. Figure 2.3 also explains why mid-northern observers are oriented towards higher galactic longitudes than mid-southern observers. Therefore mid-northern observers have a better view on the anti-center of the Milky Way and mid-southern observers have a better view of the center of the Milky Way. An observer who wants to see the whole Milky Way will need to travel across Earth's equator.

The stars that make up the constellations are also a part of the Milky Way. From our Earth-based view, the constellations give the impression of being related to specific seasons. We are used to linking Orion and Auriga with winter, Leo with spring, Sagittarius with summer, and Pegasus with autumn. Figure 2.4 shows what constellations are linked to what part of the Milky Way. The constellation of Sagittarius shows us the center of the Milky Way. The constellation of Cygnus points towards the incurving spiral arm where we belong, etc. If we want to have a look at the Orion Spur, the part of the spiral arm the Sun passes through, we should turn our view away from the center of the Milky Way. From the constellations of Cygnus through Auriga to Puppis is where our spiral arm is located.

Remember that the whole Milky Way cannot be seen from a mid-northern latitude on Earth, nor can all the constellations. The reason is that our natural horizon blocks our view to the south. Figure 2.5 shows what the hidden sectors are for a mid-northern observer.

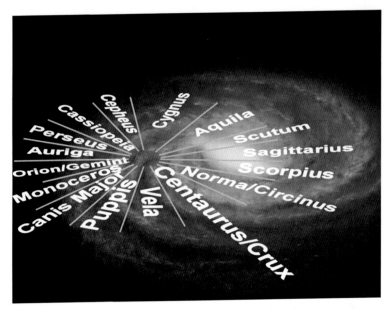

Fig. 2.4 From our viewpoint, the Milky Way surrounds us. Each of these constellations offer us a view on a particular section of the Milky Way

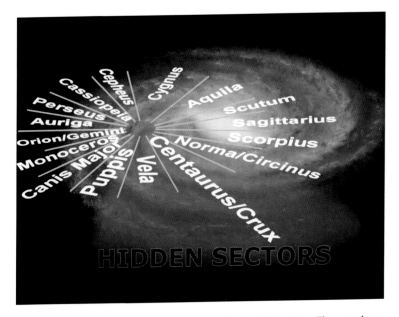

Fig. 2.5 These are the hidden sectors of the Milky Way. The northern observer's view of them is blocked by the natural horizon of the observer

The Changing Orientation of Earth

We share with our ancestors the stars and the constellations. The appearance of the most visible stars and the shapes of the constellations have not changed to a great extent during the last 6,000 years. But that's about it. The seasonal sky maps of this book would have been completely useless in the time of, let's say, the ancient Egyptians. How can that be?

The reason for this 'inconvenience' is called the precession of the equinoxes. Precession is a difficult word to explain, so let's start with the equinoxes. Literally, equinox means that day and night are both exactly 12 h long. Because the rotation axis of Earth is tilted at 23½° in relation to its path around the Sun, we have the seasons. During our summer, the northern side of Earth is pointed towards the Sun. That's why we have longer days and shorter nights. During winter, the southern side of Earth is pointed towards the Sun (see also Fig. 1.2).

Our ancestors already noticed that the Sun follows a yearly path among the stars. This path is a great circle in the sky called the ecliptic. If the axis of rotation of Earth would have been perpendicular to its orbit around the Sun, the ecliptic would be equal to the celestial equator. But because of the tilt of Earth's rotational axis, the great circle of the ecliptic is tilted with regards to the great circle of the celestial equator.

There are four important moments in a year that indicate a particular position of the Sun on the ecliptic. The winter solstice indicates the moment when the Sun arrives at its most southerly point in the sky, in the constellation of Sagittarius. For observers north of the equator, this will be the longest night of the year. The winter solstice is followed by the vernal equinox, which indicates the moment when day and night are exactly 12 h long. At that moment, the Sun is positioned perpendicular to the Earth's equator, and thus exactly at the celestial equator. If we were to project the Sun's position among the stars onto the ecliptic, we would arrive in the constellation of Pisces. This point, where the ecliptic cuts the celestial equator in the constellation of Pisces, is called the vernal point. The summer solstice indicates the moment when the Sun arrives at its most northerly point on the ecliptic, in the constellation of Taurus. Observers north of the equator then experience the shortest night of the year.

The last important moment is the autumnal equinox, which indicates the moment that day and night are exactly 12 h long again. The Sun is then in the constellation of Virgo, where the ecliptic cuts the celestial equator again. Figure 2.3 illustrates the autumnal point in relation to the galactic plane.

These four important moments in a year announce the start of a new season. It has always been vital in the history of humankind to keep track of these key moments of the year. Our ancestors tried to grab hold of these 'dates' with the seasonal apparition of the constellations. There is the one great example of the first morning apparition of Sirius, which for the Egyptians more than 3,000 years ago announced the annual flood of the Nile. The Egyptians believed that Sirius was responsible for the flooding. Today we know that the rainy season causes the swelling of the Nile.

So, in short, the equinoxes indicate the cross sections of the ecliptic with the celestial equator. These are two distinct points that can be located in the sky. The precession of the equinoxes indicates a slow shift to the west of the ecliptic with regards to the stars. This was discovered by the great Greek astronomer Hipparchus. The reason for this shift is the changing orientation of the rotation axis of Earth, due to the mutual gravitational forces exercised by the Sun and the Moon. We are all familiar with this phenomenon. It's the same wobble a spinning top exhibits when it tries to keep its balance.

The rotation axis of Earth describes a circle as illustrated by Fig. 2.6. For a backyard astronomer, this motion is negligible. Historians and cartographers however are more concerned about the precession. Celestial cartographers have to deal with the so-called equatorial coordinate system. In this system, the longitude of celestial objects (which is then named 'Right Ascension') is measured from the vernal point, and the latitude (called 'Declination') is measured from the celestial equator.

Unfortunately, sky atlases are only relevant for a certain epoch because the equatorial coordinate grid slowly drifts. Today's atlases are based on the position of Earth's rotational axis in the year 2,000 A.D. Historians, too, should take into account that the north celestial pole and the vernal point changes with time. This implies that many years ago, the local horizons on Earth differed from today. Figure 2.7 shows what the consequences of precession are.

For a mid-northern observer, the center of the Milky way does not rise very high above the southern horizon, due to the orientation of Earth in space.

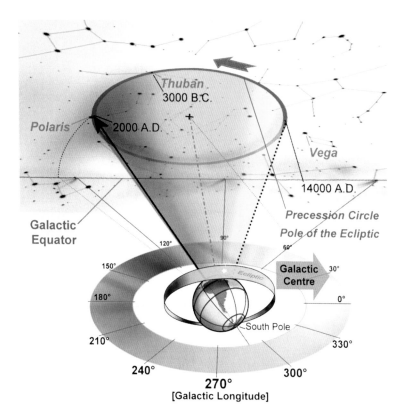

Fig. 2.6 The rotational axis of Earth traces a precession circle among the stars in about 26,000 years. About 3,000 years ago, the Pole Star was Thuban, in the constellation of Draco. In 14,000 A.D., Vega will become the new Pole Star

This is shown in the left panel of Fig. 2.7. The constellation of Scorpius hovers just above the horizon. The ecliptic also reaches its lowest point in the sky, indicating that these constellations and this region of the Milky Way are visible during the summer nights.

You might think that this is too obvious to be mentioned, but now take a look at the right panel. This panel shows the horizon from that same location in the year 14,000 A.D. Scorpius now rises much higher in the sky. Take note of the constellations that used to lie below the observer's horizon. The ecliptic now reaches its highest point in the sky. This implies that these

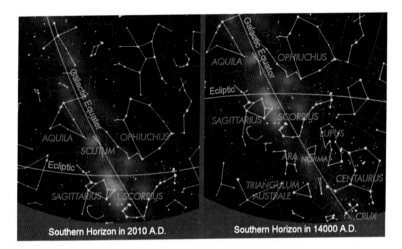

Fig. 2.7 The *left panel* shows today's southern horizon for a mid-northern observer. The *right panel* shows what happens when Earth's rotational axis is tilted in the other direction, like in the year 14,000 A.D. In this example we assume that there is no continental drift

constellations and this part of the Milky Way are visible during – winter! The slow wobble of Earth, which we call precession, determines our nighttime window to the universe as well as which constellations are visible during what season. Our seasons are dictated by the orientation of Earth. The seasonal appearance of the constellations is only a mere consequence of it.

The Structure of Our Summer Milky Way as Seen with the Naked Eye

In the previous chapter, we already discussed the benefits of binoculars for deep-sky observations. Now, it might come as a surprise to you, but on this first tour through the heavens, we will only use our bare eyes. Don't underestimate the power of your dark-adapted eyes. Many features of the Milky Way can be perfectly discerned with the naked eye from a dark location. The seasoned observer will even recognize several deep-sky objects without the use of any optical aid.

Fig. 2.8 The summer Milky Way is the brightest part of our home galaxy. The Milky Way is represented by the glowing band of light that arcs through the sky from *left* to *right*. South is at the *right* of the picture, where you can see the constellation of Sagittarius. The Cygnus star cloud shines brightly in the *middle* of the picture. Cassiopeia is at the *left side* of the frame

Our Sun is situated within the disk of our galaxy. It's not a favored position to tell the different structures of the Milky Way. But once we understand what we are looking at, our view on our galaxy looks totally different. The best visible part of the Milky Way is displayed during the summer nights, when Scorpius and Sagittarius rule the southern sky. Figure 2.8 is a picture of the summer Milky Way.

Figure 2.9 represents a sketch of the Milky Way, based on a very careful naked-eye observation. Just like with every deep-sky observation, the details don't pop into view at first glance. Only with patience and a mix of direct and averted vision can such an observation be made. Bear in mind that your view will depend on the local light pollution as well as on your latitude and of course on your experience.

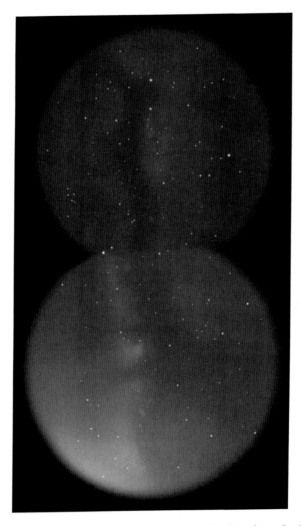

Fig. 2.9 A naked-eye sketch of the summer Milky Way from Sagittarius near the *bottom* to Cygnus at the *top* of the sketch. For this sketch, two observations were made with use of the sketching template shown in Fig. 1.26. One observation was centered upon the Scutum star cloud. The other observation was focused on the Cygnus star cloud. The sketch, which was made at a mid-northern latitude, offers a very wide view. It covers the galactic equator from longitude 0° to 90°

The glare in the lower left of the sketch, under the Sagittarius Teapot, is produced by a streetlight. You can recognize several constellations (see also Fig. 1.5). The upper circle of the sketch is filled with the cross of the Swan, Cygnus, with bright Vega at its lower left side. Under Cygnus, near the top of the lower circle, flies the Eagle, Aquila. Below the Eagle shines the Shield, Scutum, and at the bottom of the sketch boils the Teapot figure that represents the Archer, Sagittarius.

Since Fig. 2.9 is a fairly wide panorama of the summer Milky Way, a lot of details are squeezed into a tiny area. Novices, who are not used to the appearance of our galaxy, might find it very hard to initially get their bearings. For them, included are two extra, labeled views: Fig. 2.10 has the constellations and the Summer Triangle marked upon the glowing band of the Milky Way, and Fig. 2.11 has the most prominent galactic features indicated.

The Milky Way can be thought of as the hot vapor that comes out of the spout of the Sagittarius's Teapot. The cloud right above the spout of the Teapot is the large Sagittarius star cloud. It represents the central bulge of the Milky Way, at 30,000 ly from us! Here is where the central hub of our galaxy resides. Remember that this region is heavily obscured by the interstellar matter of several spiral arms. There is a small cloudlet, visible with the naked eye, at the right of the large Sagittarius star cloud. This is the Lagoon Nebula, called Messier 8 (M8). This deep-sky object lies in the next interior spiral arm, the Sagittarius Arm, at 5,000 ly from us. Above M8, there is another, bar-shaped cloud, the small Sagittarius star cloud, called Messier 24 (M24). Here is one of the windows in the interstellar matter that allows us to see deeper in the disk towards the interior of our galaxy. M24 is believed to be a part of the Norma Arm, at around 15,000 ly from us.

The stars that make up the teapot figure of the Archer are between 60 and 600 ly away. They are in fact residents of our own spiral arm, the Orion Spur, through which we look to see the far interior of our galaxy. Our Sun is located near the inner rim of the Orion Spur. Therefore, if we want to see the structure of our own spiral arm, we should turn our gaze away from Sagittarius and look at the constellations in the opposite direction.

Another tracer of an important Milky Way structure is the Scutum star cloud, centered in the lower circle of the sketch. The Scutum star cloud

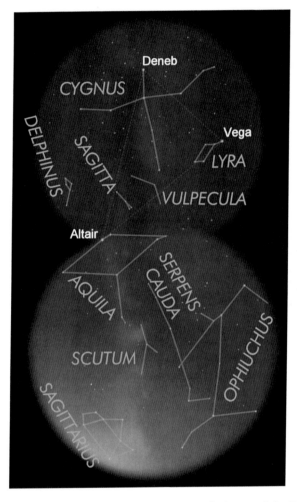

Fig. 2.10 The summer Milky Way with the constellations and the Summer Triangle formed by Deneb, Vega and Altair. The Summer Triangle shows the first stars that can be seen at dusk during summer

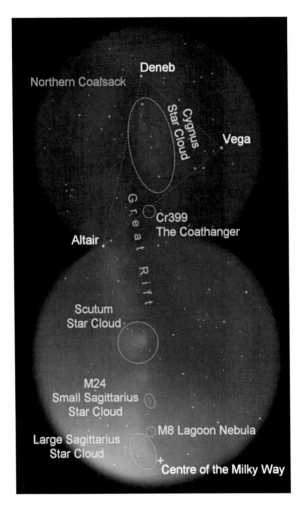

Fig. 2.11 These prominent galaxy features are visible with the naked eye under a dark sky

represents the incurving arc of the Scutum-Centaurus Arm, at a distance of 12,000 ly. This, too, is a window in the interstellar matter. The next bright Milky Way structure is located in the middle of the upper circle of the sketch: the Cygnus star cloud. The Cygnus star cloud is located at 90° longitude; thus its direction forms a straight angle with the direction of the large Sagittarius star cloud.

Fig. 2.12 A picture of the summer Milky Way from the tail of Scorpius near the horizon up to Aquila at the top of the picture. Don't expect to be able to see all these details with your naked eye. However, these glorious Milky Way features are well within reach of a pair of binoculars

Now we look deep into the cross-section of our own spiral arm, the Orion Spur. You can think of the Cygnus star cloud as the direction in which the Sun rotates around the center of our galaxy. The most spectacular and largest feature is without doubt the Great Rift. Don't seek a bright structure, though, as the Great Rift shines with – darkness! As shown in the sketch, the Milky Way is split into two parallel sections, by the obscuring masses of the Great Rift. We can only see its presence by its silhouette in front of the gleaming galactic disk. The dark dust clouds of which the Great Rift is built up are relatively close to us. The Great Rift, which is estimated to be about 300 ly away, is thus a feature of our own spiral arm. It is in such dust clouds that stars can form. These clouds of molecular dust are gigantic. The total mass of the Great Rift is equal to about one million solar masses. Be aware that the silhouette of the Great Rift can only be seen when the sky is reasonably dark.

You see that with the help of your mind's eye, the Milky Way looks totally different. The next time that you observe the summer Milky Way, you will have a deeper impression of our galaxy. And for all this celestial beauty, you don't even need any optical aid!

Photographs of constellations typically show more detail than the naked eye can see (Fig. 2.12). These photographs become very representative when using a pair of binoculars. The view of our galaxy is simply stunning when you sweep along the celestial river with a pair of binoculars. Figure 2.11 gives you an impression of the glorious celestial treasures that are awaiting you with a pair of binoculars. It is an interesting exercise to compare Fig. 2.11 with Fig. 2.9. This comparison shows how different the dark-adapted eye functions, compared to a camera. A camera is capable of recording fainter stars. It also captures the brighter galaxy clouds with greater detail, hence the mottled appearance of the Milky Way. The naked eye sees fewer stars and shows less resolution. However, thanks to the greater sensitivity of averted vision, the human eye is capable of detecting the general contours of the Great Rift, which don't show up very well in the picture.

CHAPTER 3

JANUARY

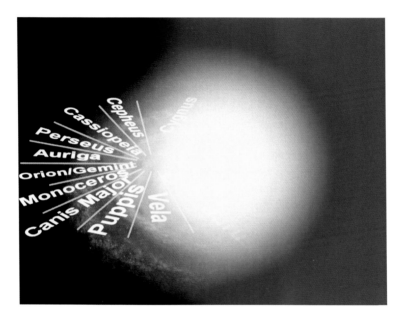

Fig. 3.1 Our nighttime window during the winter solstice is pointed towards the outer rim of our side of the galaxy. The Sun passes in front of the center of the Milky Way. That part of the Milky Way is hidden by bright daylight

Clear winter nights are bitter cold. And yet, they are incredibly beautiful. Winter nights display many fascinating objects for observers. So bundle up and grab your binoculars!

In January, the Sun passes through Sagittarius and Capricornus. It's still very close to the direction of the center of our galaxy. Therefore our best nighttime window is pointed towards the anti-center of the Milky Way, or the outer rim of our side of the galaxy (Fig. 3.1). The major part of the Orion Spur,

R. De Laet, *The Casual Sky Observer's Guide*, Astronomer's Pocket Field Guide, DOI 10.1007/978-1-4614-0595-5_3, © Springer Science+Business Media, LLC 2012

our local spiral arm, also lies in the direction of the anti-center. This month, we'll visit the constellations of Auriga, Gemini, Orion and Canis Major.

The objects for this month's tour are located so to speak in our own backyard, the environment of the Orion Spur. Figure 3.2 shows where exactly the deep-sky objects are positioned. They all are inhabitants of the galactic disk. During this month around 22:00 local time, Auriga passes through the zenith, to its SE, followed by Gemini, Orion shines high in the South, and Canis Major lingers above the SE horizon.

As you can see in Fig. 1.9, the winter Milky Way is highly inclined compared to our southern horizon. The constellations of interest are easy to identify with the use of Fig. 1.9. Auriga, the Charioteer, is highlighted by its mag zero beacon, Capella. You simply can't miss Capella. It's the brightest star overhead. You should be able to identify Auriga as the 15° wide pentagon of which Capella is the NW corner-stone. Gemini is easily identified by its two bright stars, Castor and Pollux. Orion is the remarkable figure of seven stars that dominates the southern sky. Some people see an hourglass, others call it a butterfly on its side. Orion's waist is defined by three equally bright stars, often called the Belt of Orion. It's only 3°

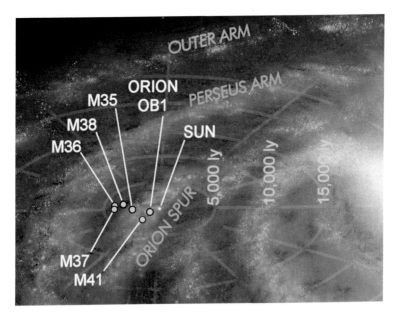

Fig. 3.2 The objects for this month's tour are located near or in our own spiral arm

wide. Orion's shoulders are marked by two bright stars. His left shoulder is a star called Betelgeuse. The two remaining stars under Orion's waist represent his feet. The star that marks Orion's right foot is Rigel. Canis Major is dominated by the brightest star in the sky, Sirius, shining at mag −1.5! Follow the line of the three stars of Orion's waist to the SE and you bump into scorching Sirius. The shape of Canis Major is easier to detect later on the night, or later in the month, when the constellation is at its highest.

Messier 36, 37 and 38

Our first deep-sky objects of the month are located in the constellation of Auriga. Like you can see on the finder chart (Fig. 3.3), Auriga lies on the galactic equator, represented on the chart by the dashed line. On a dark night

Fig. 3.3 Finder chart for M36, M37 and M38

you can see how the Milky Way passes through the pentagon of Auriga. Because the view through this part of our spiral arm is not obscured by nearby dust clouds, Auriga harbors many open clusters. The most famous Auriga open clusters are the Messier trio M36, 37 and 38.[1] They are a fine sight in our binoculars, and offer an interesting comparison. You might like to think of Auriga as a jewel box that's been thrown in the celestial river. The strong current washes the Messier trio out of the box.

Let's start with Messier 37, which has already left the jewel box. This open cluster is the easiest to locate. Our signposts are Theta Aurigae and Beta Tauri, which lie on the border with Taurus, the Bull. Just point your binoculars in between these two stars. M37 should appear in your field of view as a hazy patch of light. With the use of your averted vision, the size of M37 could grow to 20', a little smaller than the disk of the full Moon. If your sky is not reasonably dark, the cluster will appear smaller. With a pair of 8×56 binoculars, the cluster remains unresolved. Only the brightest cluster member, its mag 9.2 lucida, could be discerned from the background glow of the cluster. No wonder that Messier included this cluster in his catalog. M37 could easily be mistaken for a comet (Fig. 3.4).

M37 is a bright and rich open cluster, shining at mag 5.6. It lies 4,500 ly away and measures 33 ly across. M37 consists of about 2,000 stars. Because the cluster is 350 million years old, it has lost already its brightest O-stars. That's the reason why binoculars cannot resolve this cluster. There are more than 35 red giants[2] and a remarkable number of white dwarfs[3] in this cluster. Its integrated absolute magnitude is −5.7. A star like our Sun would shine with an apparent magnitude of 15.5 from within the cluster. It would thus remain invisible in our binoculars.

---

[1] Charles Messier was an eighteenth century French comet hunter, who published a catalog of deep-sky objects that were not comets. This list is often called the 110 Messier Objects.

[2] Red giants are older stars that have exhausted their hydrogen supply, resulting in a contracting core that heats up the outer layers that expand. Our Sun, too, will end as a red giant. The presence of red giants in a cluster is a measure for the age of a cluster.

[3] White dwarfs are the final evolutionary stage of stars that became red giants. They have ejected their outer shells in a planetary nebula. What remains is a very dense core of inert matter. A white dwarf with the mass of our Sun would be the size of Earth.

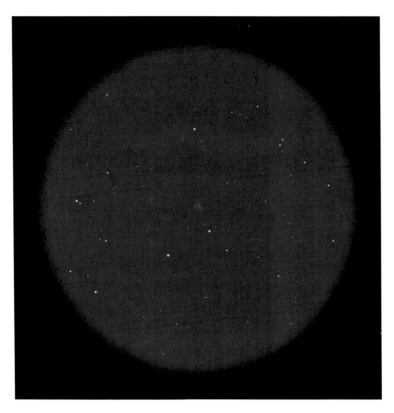

Fig. 3.4 Messier 37, sketched after being observed with a pair of 8×56 binoculars

From M37, it's a small step to M36 and M38. As you can see in Fig. 3.3, both clusters are only one binocular field away from M37. Just move from M37 to the NE and both open clusters should appear in your sight. If your binoculars have a very wide field of view, the three clusters might fit in the same fov. Another way to find M38 is to center your view in between Theta and Iota. There you should see M38 in the middle of your sight. M36 is only 2°19' to the SE of M38.

M36, at a distance of 4,300 ly, is the smallest of the trio. It only measures 15 ly across. About 180 stars are counted in this cluster, a low number compared to the 2,000 members of M37. M36 has an integrated absolute magnitude of −5.5 while its apparent magnitude is 6.0.

As you can see on Fig. 3.2 M36 and M37 are neighbors. They are perhaps 400–500 ly separated from each other and have about the same absolute brightness. But how different they look! In 8×56 binoculars, you could easily resolve ten stars of mag 9. In fact, M36 is the brightest and shows the best resolution of the trio. The reason is that M36's brightest stars are very hot blue giant B2 an B3 stars. Therefore, M36 is believed to be very young, with an age of about 42 million years. Given its age and location near the Orion Spur, M36 might be a true tracer of our spiral arm. The proper luminosity of M36's blue giants is enormous, compared to our Sun, which would merely shine at mag 15.4 when placed in this cluster. M36 is similar to M45 (see the December objects) in appearance, only its ten times further away.

M38 seems to be the largest of the trio.[4] It does not resolve completely, but it displays a remarkable spread out π-shape if you observe carefully. The cluster's lucida is a mag 7.9 resident of the eastern leg of the π-shape. M38's appearance can be explained by its age and distance. M38 is a more evolved cluster than M36, with an age of 360 million years. Its distance is estimated at 3,500 ly. With an apparent brightness of mag 6.4, M38 is the dimmest open cluster of the trio. It has a true physical size of 15 ly.

When the sky is very dark, you can try your luck finding NGC 1907. It's a 5 arcmin-wide open cluster that can be found 30' south of M38. Its apparent magnitude is 8.2, so use your averted vision at its best to detect this 4,300 ly distant open cluster. As shown in the sketch (see Fig. 3.5), NGC 1907 appears as a faint smudge of light under M38. It's a nice exercise in training your galactic perspective: M38 lies in the foreground at 3,500 ly while M38 and NGC 1907 decorate the background at 4,300 ly.

The Messier trio of open clusters are still very young, considering the age of the cosmos. But we find it hard to get grip on astronomical ages.

---

[4] M38 appears large because it has bright members spread out over its entire area. Under dark skies, M37 will appear larger.

Fig. 3.5 Sketch of M36 and M38 as seen with a pair of 8 × 56 binoculars

A more striking timescale is the cosmic calendar. The cosmic calendar represents the complete lifetime of the universe (15 billion years) in one calendar year. A year is a more comprehensible timescale for us to master. Let us say that the Big Bang, the beginning of cosmic time, happened on the first of January, and today is the last second of 31st of December of the cosmic calendar. The Milky Way formed on May 1. The Solar System formed on September 9. Messier 38 dates from December 21, when the first vertebrate land animals populated Earth.

The cosmic calendar is a great tool in understanding the gargantuan ages of astronomical objects. How about us humans then? Well, we don't score high on the cosmic calendar. Humans have only existed since 10:30 p.m. on

New Year's Eve. Things get even worse when we consider our individual lifespan. An average human life only lasts for about 0.15 s. So, we better move on to the next object.

Messier 35

The stars of Messier 35 bring us a little closer to home. This open cluster resides in the constellation of Gemini, the Twins. Locating M35 is fairly easy. The Twins are represented in the sky with one foot each. M35 lies at Castor's foot. It's sometimes called the Shoe Buckle cluster. It's a small star hop from μ Gem to η Gem. With η in the SE corner of your binocular field, M35 shines brightly at mag 5.1 in the center of your view (Fig. 3.6).

Messier 35 is a very rewarding object in small binoculars (Fig. 3.7). You could partially resolve this sparkling cluster with your 8 × 56 binoculars. Four stars are visible with direct vision, while a total of 14 stars could be picked out of the glow with averted vision. A few dark lanes are visible as well.

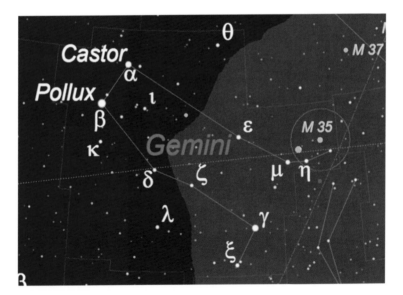

Fig. 3.6 Finder chart for Messier 35, located near the northern foot of the Twins

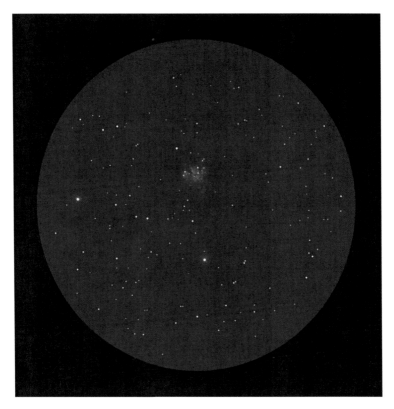

Fig. 3.7 Messier 35 as seen in a pair of 8×56 binoculars. The bright star near the left edge is η Gem

Figure 3.2 shows that M35 is closer to us than the Auriga trio of Messier objects. Its distance is estimated at 2,700 ly. The cluster's true diameter is 22 ly, which explains its apparent diameter of 28′. Messier 35 formed possibly 150 million years ago and consists of over 2,700 stars. Clusters like these are dominated by the brilliance of their hot giant stars. From within this cluster, our Sun would have a paltry apparent magnitude of 14.4.

Once you have carefully studied M35 with your binoculars, why not try your averted vision without optical aid? With Gemini high overhead, the cluster is possibly a naked-eye object, providing your sky is reasonably dark enough. The naked-eye sighting of M35 is an excellent way to hone your observing skills, as it is a measure of the quality of your observing site.

Collinder 70

We arrive still closer to home with our next stop, the constellation of Orion. The eye-catching Belt stars, ζ, ε, δ, together with Saiph, Rigel, λ and ι all belong to the Orion OB1 association (see Fig. 3.8). They are the hot O- and B-stars emerging from a immense molecular cloud, the Orion molecular cloud complex, which envelops a substantial part of the constellation.

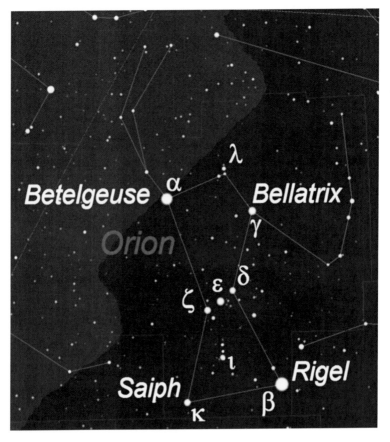

Fig. 3.8 The constellation of Orion

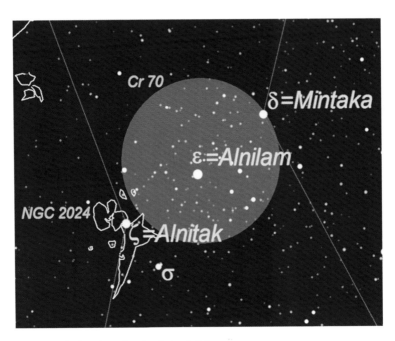

Fig. 3.9 Finder chart for the Belt of Orion

The Orion OB1 association is a real treat for the naked eye and binocular observer. In binoculars, Betelgeuse's orange color offers a nice contrast with Bellatrix or the blue-white Belt stars.

The Belt stars are part of a rich star field that bears the name Collinder 70 or Cr70.[5] It is a 2° wide open cluster that is best observed at the lowest possible magnification. Cr70 lies 1,200 ly away and is only four to five million years old (Fig. 3.9).

The Belt Cluster is a magnificent object in any pair of binoculars. Just aim your gaze at Alnilam and you have Cr70 in your fov. The three Belt stars are very powerful silver-blue giants. They just overpower the rest of the star field with their luminance of 60,000 Suns each. Scores of fainter Stars form

---

[5] Per Collinder was a twentieth century Swedish astronomer who published a catalogue of open clusters.

Fig. 3.10 The Belt Cluster Cr70 with NGC 2024, observed with 8×56 binoculars

a serpentine snake around the Belt stars. Our Sun would pale to near nothing compared to any of the observed stars within Cr70, as it would shine with an apparent magnitude of 12.6.

Cr70 has blown away most of its original gas and dust cloud. There is still some nebulosity visible in this area. The astute observer can try his or her luck on the emission nebula NGC 2024, 15′ east of Alnitak. NGC 2024 is a part of the Orion molecular cloud in which star formation is taking place. The radiation of the hottest giant stars is powerful enough to ionize the surrounding gas, which starts to glow. In the case of NGC 2024, it's a complete embedded cluster that's responsible for the nebula's glow. There might be more than hundred stars hiding inside the obscure cover of NGC 2024, which measures 11 ly across. The nebula is called the Flame Nebula, because photographs show a burning bush of blazing light. But NGC 2024 also bears the name 'the Ghost of Alnitak,' which may be a more appropriate name from an observer's point of view. The nebula itself has an integrated magnitude of 7.2, which is reasonably bright. Alas its close proximity to Alnitak makes NGC 2024 an incredibly difficult object to observe. Alnitak's glare interferes with our night vision. Telescopic observers can crank up the magnification and keep Alnitak out of the fov. With our binoculars, there is no way around Alnitak unless you could hide the Belt of Orion behind a nearby tree or building. On the darkest of nights, it is feasible to discern the Ghost of Alnitak with averted vision, as Fig. 3.10 shows.

The Sword of Orion

There is a row of faint stars dangling under Orion's Belt. It is often called the Sword of Orion, as the Greeks imagined Orion as a warrior. The Sword of Orion is the absolute winter showpiece of the heavens. Here is where the most famous and brightest of all the nebulae resides: The Orion Nebula, M42. Messier 42 fares better in telescopes, but binoculars are great instruments to display the Orion Nebula in its natural habitat of the Sword.

Figure 3.11 shows how easy it is to find the Sword of Orion. It lies in the middle of Orion's underbelly, just one binocular field south of Cr70, the Belt Cluster.

This region of the Orion OB1 association is filled with very bright stars. Many of these stars are very young giants or supergiants. The brightest star of the scene is ι (Iota) Orionis. It's an O9 giant with a surface temperature of 31,500 K that is maybe only 500,000 years old . Just 8 min SW of ι Orion is the double star Struve 747. Its components are both B-stars and only 36″ apart, but clearly split at 7×. Struve 747 may look a little elongated in the sketch, due to the lower resolution of the rendering. It appears that ι Orionis could be the lucida of a little poor open cluster, of which Struve 747

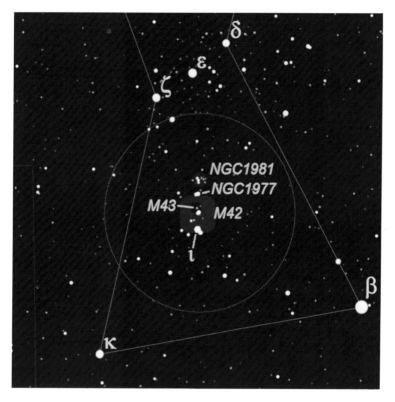

Fig. 3.11 Finder chart for the Sword of Orion

is also a member. The showpiece of the scene is M42, the Orion Nebula. A pair of 8 × 56 binoculars reveal two stars in the middle of M42: θ1 Orionis and θ2 Orionis. The latter is accompanied by two fainter stars to the east. The θ1 lies in the brightest part of M42.

The heart of the Orion Nebula is extremely bright. It can be seen with direct vision, also from an urban location. With patience and averted vision, a larger part of the nebula can be witnessed in a pair of binoculars. Several faint stars can be discovered in the fading glow, too. The western part of the nebula is the largest and the most obvious 'wing' of M42. Its northern border brightens towards the edge and then seems to be sharply cut away, as if a dark nebula separates M42 from the fainter M43 a few minutes to the north of the θ-pair.

M43 looks like a faint star embedded in a misty glow. The dark nebula also curves S as if it wants to separate the θ stars as well. This dark intrusion into the Orion Nebula is called the Fish's Mouth. The southern 'wing' is a very diffuse feature. It is the thick, long filament, known from the photos, that point towards ι Orionis. With averted vision, this filament can be seen with 8×56 binoculars. The appearance of the both wings together with the θ-pair reminds some of a flying bat.

The Orion Nebula is one of the few objects that shows true structure in binoculars. M42 is sure to become a favorite object that you will want to revisit often. What we see is a cavity within a molecular cloud that fluoresces due to the ultraviolet radiation of the hot stars from within. Many of the stars in the Sword of Orion are close double or multiple stars. With higher magnification θ1 Orionis would split up into four components, called the Trapezium. This quartet of hot O- and B-stars is extremely young. They are responsible for the gossamer glow of the Orion Nebula. Their strong radiation and stellar wind pushes away the remaining interstellar dust and gas. It is believed that these powerful stars are in part responsible for the compression and condensation of nearby molecular clouds into birthplaces for other stars. M42 is located at a distance of 1350 ly and measures 30 ly across.

The mag 7 star a few minutes north of the Fish's Mouth is embedded in a very diffuse halo. This is M43, fluoresced by the central star that is an O-star also. M43 is the companion of M42, separated by the dark dust lane as seen in the sketch (see Fig. 3.12). Because of M43's low surface brightness, it will only show up on the clearest of nights.

At 30' north of M42, the extremely faint glow of the reflection nebula NGC 1977 shows up around 42 and 45 Orionis. Both stars are cataloged as early B-stars that haven't developed sufficient radiation to fluoresce NGC 1977. A total of three stars can be seen within the nebula.

Fig. 3.12 The Sword of Orion as seen with a pair of 8 × 56 binoculars

The final object is the young and widespread open cluster NGC 1981, north of NGC 1977. NGC 1981 marks the northern end of Orion's Sword and has a true physical size of 11 ly. A pair of 8 × 56 binoculars show a total of about ten medium to faint cluster members in an area as large as the full Moon.

The above mentioned objects are all members of the OB1 association, at a distance of about 1,300–1,500 ly. The Sword of Orion offers the binocular observer a unique combination of glowing nebulae and sparkling star clusters in one single field of view. On a clear moonless night, the Sword of Orion is also a fascinating naked-eye sight. See if you can resolve the Sword into three or four fuzzy stars. The OB1 association is a marvelous section of our own spiral arm. It displays such cosmic splendor that it is worth every minute of suffering through the freezing cold of clear winter nights.

Messier 41

Our last destination for this month is the open cluster of M41 in the constellation of Canis Major, the Larger Dog. Canis Major is a constellation near the galactic equator that in contrast with Orion is not obscured by molecular clouds. Thus Canis Major offers us a free view into the deeper parts of our local spiral arm. Here we find several interesting star fields and clusters.

Canis Major is also the home of the brightest of all stars, scorching Sirius. Sirius is pointed to by the Belt stars of Orion. If you follow the direction of the three Belt stars to the SE, you arrive at the mag −1.46 Sirius. It's a dazzling star to behold in your binoculars, as it will ruin your night vision. We'll take a closer look at Sirius in the next chapter. For now, Sirius is our signpost on this tour. If you move your binoculars 4° to the south until Sirius is out of sight, another sparkling celestial jewel comes into view: Messier 41. Take your time to adjust your eyes again to the darkness of the fov. M41's manifestation is a lot subtler than Sirius's prominent scintillation (Fig. 3.13).

A pair of 8 × 56 binoculars show 10- to 12-min stars irregularly spread over an area of 35′. There definitely is a background glow of unresolved stars. The mag 6.3 lucida of the cluster is located near its center. It is a K3 red giant. The brightest stars of this cluster are already evolved into G- and M-giants. M41 is a fine cluster that deserves to be observed with patience. A little imagination brings the figure of the hanging bat to mind. It's a comparison used by Stephen James O'Meara in his book, *The Messier Objects*. The bat seems to reach for the mag 6 star, 12 CMa, the brightest star in the field, SE of the cluster. 12 CMa is a B7 foreground star at 670 ly, while M41

Fig. 3.13 Finder chart for Messier 41

lies at 2,300 ly from us. The 26 ly wide open cluster is probably 280 million years old. Messier 41 has an apparent magnitude of 4.5, which makes it an interesting naked-eye object. Unfortunately, the cluster does not rise very high above the southern horizon; otherwise it would gleam more. The cluster is best viewed during January and February evenings. Its low apparition above the horizon makes it very susceptible to being obscured by light pollution and atmospheric moisture (Fig. 3.14).

Fig. 3.14 Messier 41 sketched with a 4 in. refractor at ×19. The field is 2.5 wide

FEBRUARY

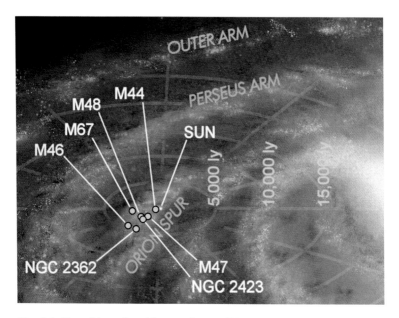

Fig. 4.1 The objects for this month are all located within our own spiral arm, the Orion Spur

This time of the year, the Sun moves through Capricornus and Aquarius, on its route to the Vernal Equinox. February is a perfect time to discover the deep-sky treasures of the more southern Milky Way constellations of winter. After dusk, Sirius shines brightly in the south. It's the final call to the fascinating destinations of our own galaxy. That's right, next month, we'll depart from the Milky Way, but not before we've visited the sparkling clusters within the deepest stretch of our local spiral arm, the Orion Spur (Fig. 4.1).

R. De Laet, *The Casual Sky Observer's Guide*, Astronomer's Pocket Field Guide, DOI 10.1007/978-1-4614-0595-5_4, © Springer Science+Business Media, LLC 2012

Last month we had the opportunity to visit the number one deep-sky object of winter, the Orion Nebula. This month we'll sail on to lesser known but equally beautiful destinations. Our February tour brings us from Canis Major, the Larger Dog, to Puppis, the Stern, followed by Monoceros, the Unicorn, and Hydra, the Water Snake, up to Cancer, the Crab. All these exotic places are located in or near the celestial river of the Milky Way. These constellations are less obscured by giant molecular clouds than Orion is, and they offer us a clear view into the structure of the Orion Spur.

Sirius, the Dog Star

Sirius is, next to our Sun, the brightest star in the sky. Only Jupiter, Venus and the Moon can beat Sirius in brightness. If you have toured the January objects, you already know which star Sirius is. If not, check Fig. 3.13. Sirius is pointed to by the Belt stars of Orion. Just follow the three stars of Orion's waist to the SE into the constellation of Canis Major, the Larger Dog, where you'll encounter the king of Suns, shining magnificently at a magnitude of −1.46.[1]

Sirius is also the closest naked-eye star for mid-northern observers. But even long before its true distance was understood, this star played an important role in human history. For the ancient Egyptians, the heliacal rising[2] of Sirius heralded the annual flooding of the Nile, which was a fundamental life-giving occurrence in an agricultural civilization. No wonder that they celebrated this event as the New Year's Day of their civil calendar. The Greek name for Sirius was Seirios,[3] referring to the stars immense brightness and supposed heat. The ancient Greeks credited Sirius for the hottest days of summer, which we still call the dog days. They imagined

---

[1] Sirius is roughly 1,600 times brighter than the faintest star visible with the naked eye under pristine conditions.

[2] The heliacal rising of Sirius is the first apparition of the star in the bright dawn before the Sun rises. In the era of the Egyptians, this moment came around the summer solstice. Due to the precession, the heliacal rising of Sirius now occurs around the 10th of August.

[3] Seirios means "the scorching one."

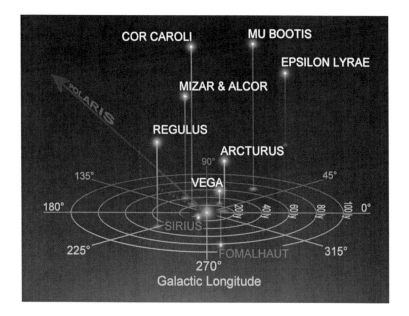

Fig. 4.2 The featured naked-eye stars in the neighborhood of our Sun. The Sun is at the center of the sketch. The concentric circles represent the galactic plane. The sketch has the same orientation in space as the galactic views of our Milky Way. The center of our galaxy is thus located in the direction of 0° longitude. Only two stars, Sirius and Fomalhaut, are located below the galactic plane

that the combined blistering heat of the Sun and Sirius, knowing both were present in the day sky during summer, was responsible for the warmest days of the year and for the diseases that came with them. Sirius was thus feared for its thought malignant influence. Of course, all of this is just a coincidence. For the ancient Polynesians, Sirius was one of the many important navigational references in the sky. Polynesian mariners used the stars to navigate between the numerous islands of the Pacific Ocean.

You can see Sirius and all the other featured naked-eye stars in Fig. 4.2. These stars are so close to the Sun that they would not properly show up on the galaxy view of Fig. 4.1. The fact that a much smaller scale was needed to draw the featured naked-eye stars in the Sun's neighborhood is another demonstration of the incredible vastness of our galaxy. The distance between individual stars is extraordinarily large. If you tried to build a scale model of Fig. 4.2 with marbles to represent the stars, it would be an impossible task. Using a 5/8″ marble to represent our Sun, Earth would be a 0.1 mm grain at a distance of 1.7 m. At that scale, Sirius would lie 929 km (577 miles) away. The center of the Milky Way would still be a whopping 3.2 million km (2 million miles) away, which is more than eight times the distance from Earth to the Moon.

There are two reasons why Sirius is the brightest star in the sky. Sirius is intrinsically bright because of its absolute magnitude of 1.45, and Sirius is only 8,6 light-years away. But what makes Sirius a really interesting star is the fact that it has a special companion. That's right, Sirius is a binary star system. The bright one is called Sirius A, and the companion is called Sirius B, or sometimes "the Pup." The companion is roughly 10,000 times dimmer than Sirius A. It has an apparent magnitude of 8.44, which normally is within reach of a pair of binoculars. But the Pup stands so close to Sirius A that it remains hidden in the glare of the latter. Only a powerful telescope is capable of revealing Sirius's companion.

Sirius A is a main sequence[4] A1 star with a surface temperature of 9,900 K and an absolute magnitude of 1.45. It has twice the mass and 1.75 times the diameter of our Sun. Sirius B is what astronomers call a white dwarf. It has a surface temperature of 24,800 K and roughly the mass of our Sun, but all its matter is crammed into a sphere as small as 0.92 the diameter of Earth. One spoonful of the Pup's substance would weight 1.7 metric tons, the equivalent of a pickup truck.

Sirius B was the first discovered and also the closest white dwarf to us. But what are white dwarfs? In short, white dwarfs, or degenerate dwarfs, are the cinders of the cosmos. They represent the final stage of a star not heavy

---

[4] A main sequence star, or dwarf star, is a stable star that burns hydrogen. The main sequence phase is the longest one in the lifetime of a star, until the hydrogen is consumed. Our Sun is a main sequence star. What happens with a star when it has exhausted all its hydrogen depends on the mass of the star. Stars with up to ten solar masses become red giants. More massive stars can explode as a supernova.

enough to explode as a supernova. All stars with a mass lower than eight solar masses are believed to become red giants once their hydrogen supply is exhausted. At the end of the red giant phase, the star will throw off its outer shell, which forms a planetary nebula. The exposed hot core forms the white dwarf. Because the core no longer produces energy, the white dwarf slowly cools until it's no longer visible.

Astronomers have concluded that the Sirius binary system is roughly 240 million years old. The companion of Sirius A was in fact once the primary star of the system. It was a hot blue B3–B5 star of about seven solar masses. It consumed its fuel rather quickly and became a red giant and finally a white dwarf after 120 million years. So if you think that Sirius appears bright, imagine how bright the Sirius star system might have shined in the past with both stars on the main sequence.

The history of Sirius teaches us a lot about the chemical evolution of our galaxy. The young universe contained only hydrogen and helium. The heavier elements were forged in the hot cores of the stars. The majority of the stars return a large amount of their enriched content back into the interstellar medium, through planetary nebulae or supernova explosions. We owe our existence to the death of other stars, that enriched the molecular cloud in which the Solar System formed. The book you are reading consists of atoms that have been created inside the hot cores of long-gone stars.

Sirius is a splendid object in any pair of binoculars (see Fig. 4.3). This star is famous for its breathtaking appearance in multiple colors. Its radiates not only in the typical blue-white color of an A1 star but often sparkles with various colors, which occurs when atmospheric turbulences act like a prism. This effect is strongest when bright stars are near the horizon and has probably been observed since antiquity. According to Craig Crossen in his book *Binocular Astronomy*, the ancient Sumerian[5] name for Sirius was "Mul Tiranna," the Rainbow Star. See for yourself if you can notice the prismatic sparkling of the Rainbow Star when it's low above your horizon.

---

[5] Sumer is the earliest known civilization, based in modern Iraq, which began in the 6th millennium B.C.

Fig. 4.3 The brightest star, Sirius, sketched after viewing with a pair of 8 × 40 binoculars. The Pup remains hidden in the glare of Sirius A. Our Sun would shine like a mag 1.9 star if placed next to dazzling Sirius

NGC 2362

One of the most interesting deep-sky objects in Canis Major is NGC 2362. Use Fig. 4.4 to locate δ, ε and η in the constellation of the Greater Dog. The three stars just fit in the same fov of your binoculars. If you then place δ in the SW of your fov, Tau CMa will be centered in the middle of your sight.

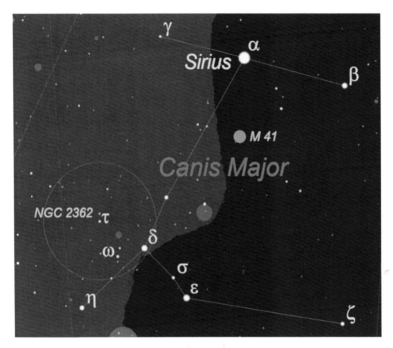

Fig. 4.4 Finder chart for NGC 2362

Here we see a small and tight open cluster of extremely young stars. The cluster is dominated by the mag 4.4 τ Canis Majoris, a blue-white hot O9-star. Around it, numerous fainter companions are gathered in an area as large as half a Moon disk. Telescopic observers have the advantage that they can use higher power to resolve the cluster. Binocular observers will find it very hard to actually see any of NGC 2362's members in the glare of its lucida. Higher power binoculars are better suited for the task. You should be able to detect NGC 2362 with a pair of 20×80 binoculars. The true nature of the cluster is easily revealed in the eyepiece of a telescope. NGC 2362 is an exciting mix of brighter and fainter stars that swirl around Tau. But the bright Tau largely outshines the fainter cluster members. That's why it's hard to maintain night vision with Tau in the picture. With averted vision, more spots of unresolved starlight light up around the cluster. How magnificent the cluster must shine when higher in the sky! (Fig. 4.5).

Fig. 4.5 A sketch of the Tau CMa Cluster, NGC 2362. A 4 in. refractor at ×63 was used to observe this young cluster. The faintest stars are of mag 12. The fov is 65′ wide, which is only one-sixth of a common binocular field of view

Tau and NGC 2362 lie deep within the Orion Spur. Their distance is estimated at 4,500 ly. The cluster's age is about five million years, which is extremely young. Many of the cluster's stars are still contracting to become steady main sequence stars. Tau itself is an extraordinary multiple-star system of possibly five components. Each of them is thought to be a heavy star of about 20 solar masses, which explains their true luminance. These massive stars are all destined to explode in a violent way in the "near" future. NGC 2362's appearance will definitely have changed within ten million years, when it has lost its hot O-stars. Our Sun, when placed in this distant and brilliant cluster, would gleam with the pale light of a mag 15.5 dwarf.

M46, M47 and NGC 2423

Our next destination is a particularly attractive trio of open clusters. They are located in the constellation of Puppis, the Stern. Unfortunately, the region of our interest is void of really bright stars. To make things easy, we'll use the constellations of Monoceros and Canis Major as a signpost for this tour. With a little help from your imagination and the finder chart of Fig. 4.6, you can see a flat triangle formed by Sirius, Kappa Orionis and Alpha Monocerotis. Try to locate this triangle in the sky without optical aid first. Then aim your binoculars at Alpha Monocerotis. Once you've located Alpha, you're almost finished. The final step is easy. Just move your sight due south of Alpha Mon, until it has left the fov. The bright clusters that have entered your eyepieces are this tour's targets.

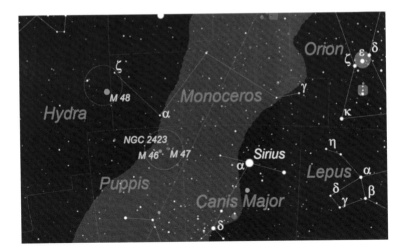

Fig. 4.6 Finder chart for M46, M47, NGC 2423 and M48

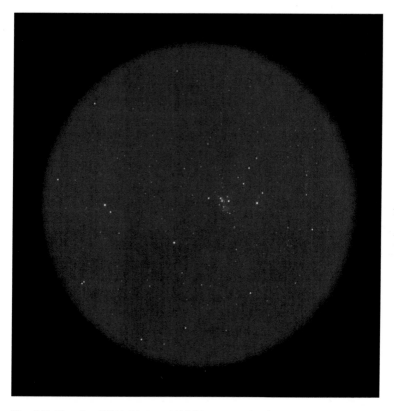

Fig. 4.7 Sketch of M46, M47 and NGC 2423, made after viewing with a pair of 8 × 56 binoculars. The fov measures 5.5° across

Your fov should look like the sketch of Fig. 4.7. Here you find a really deep field of star clusters. M47 is the brightest cluster with the best resolution. M46 lies 1½° east of M47. M46 is much harder to resolve. NGC 2423 is the small patch of light just north of M47. These marvelous clusters just happen to lie in the same line of sight. They are however not physically related to each other, as Fig. 4.1 shows.

M47 is a naked-eye object with an apparent magnitude of 4.4. It is a young cluster at a distance of 1,600 ly. It shines brightly thanks to its dozen powerful B-stars. The cluster is 15 ly wide and consists of approximately 50 stars. Its age is estimated at 30 to 100 million years. The cluster's integrated absolute magnitude is −4.4.

M46 looks completely different, although it has an integrated absolute magnitude of −4.0. This cluster is also three times as distant as M47. M46, with its over 500 stars, is one of the richest Messier clusters. It measures 25 ly across and lies at a distance of 4,500 ly. The clusters brightest members are resolvable in a pair of 8 × 56 binoculars mounted on a tripod. About 500 million years old, M46 no longer owns massive O- or B-stars. Its brightest members are A-stars and evolving red giants. The contrast between M46 and M47 is overwhelming. The latter is a coarse bunch of white-bluish O-stars, while the former is a gleaming cloud of gently glittering starlight, especially when observed with averted vision.

When the sky is clear, another cluster can be detected north of M47. If you follow a small chain of faint stars north of M47, you'll arrive at NGC 2423. This is a mag 7 open cluster at a distance of 2,500 ly. A pair of 8 × 56 binoculars show two mag 9 stars, embedded in a distinct smudge of light. This cluster is 950 million years old and is 15 ly wide.

When this trio of star clusters is observed in a pair of binoculars, the field of stars in the eyepieces looks really deep. M47, clearly resolved, lies in the front, with NGC 2423 behind it. M46 appears larger but also more distant. Try and see if you, too, can experience this wonderful sensation of true galactic depth.

Messier 48

Alpha Monocerotis was our signpost for the previous objects. It will also feature as our starting point for our star hop towards our next destination, Messier 48. From α Mon move about two binocular fields towards the NE, where you'll find the mag 4.3 ζ Mon. Figure 4.6 shows that α and ζ are only separated by 9°. You can even try to locate Zeta Mon with the naked eye. When you put Zeta near the NW edge of the field of view of your binoculars, you have just crossed the border between Monoceros and Hydra, where you find the bright and 30′ wide galactic cluster M48.

With a magnitude of 5.8, this partially resolved showpiece is even a naked-eye object under dark skies.

Messier 48 has a dozen stars that are brighter than mag 9.7, which are visible with a pair of binoculars. One's first impression of the central condensation of this open cluster might be that it looks like the body of a butterfly. The distribution of the other cluster members can be thought of as the butterfly's fluttering wings. With higher power, more stars can be distinguished out of the cluster's subtle background glow. The sketch of M48 in Fig. 4.8 shows stars up to mag 11.

Fig. 4.8 M48, observed with a 4 in. refractor at 16×. The fov is 3° wide

The cluster is thought to be 2,500 ly away, with a proper diameter of 24 ly. The true luminosity of M48 is estimated at −3.6. Its brightest members are G- to K-giants, while the hottest main sequence star is an A1-star. Messier 48 consists of about 120 stars and is believed to be roughly 500 million years old. The cluster's age could explain why it has already drifted 500 ly away from the galactic plane. On our next tour, we will visit open clusters that floated even much further away from the galactic plane.

Messier 44

Our next destination, the constellation of Cancer, lies quite a bit out of the winter Milky Way. Here we find one of the most pleasing binocular highlights of February, the Beehive (aka Messier 44 or Praesepe). However Cancer, the Crab, is a rather discrete constellation. Its brightest stars range from mag 3.5 to 4.

We can use the Milky Way constellations of Gemini and Canis Minor to locate the constellation of Cancer. Both are represented on the finder chart of Fig. 4.9. Try to locate the naked-eye quadrangle formed by

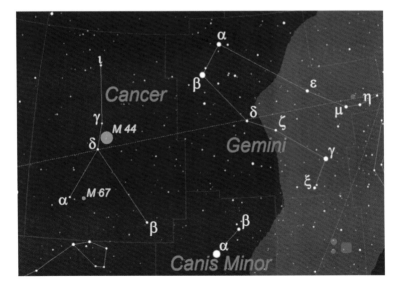

Fig. 4.9 Finder chart for Messier 44 and Messier 67

β Geminorum, γ Geminorum, α Canis Minoris and δ Cancri. Now take your binoculars and bring δ Cancri in your fov. The Beehive lies a mere 2° NW of δ Cancri. The cluster is even brighter than δ Cancri and thus is visible with the naked eye on a moonless night. As you can see on the finder chart, Cancer lies on the ecliptic. Occasionally one of the brighter planets passes through this constellation, making it harder to identify Cancer's brightest stars.

The Praesepe is one of the few deep-sky objects that was known before the invention of the telescope. Ptolemy observed the cloudy patch from within ancient Egypt, and he imagined that it was the breast of the Crab. Ancient Greeks and Romans called it Praesepe, which is Latin for the manger. Although it is possible to partially dissolve this 30'-wide cluster with the naked eye under the best circumstances, Galileo was one of the first to discover the true nature of the Praesepe with the use of a telescope. Charles Messier included the cluster in his list of comet pretenders in 1769. The name Beehive was introduced sometime in the nineteenth century. It refers to the telescopic appearance of the cluster.

With a simple pair of binoculars the Praesepe resolves into a beautiful triangular open cluster measuring more than 1° across. Near its center lies a house-shaped asterism. Several multiple stars are visible at low power. The binocular view is perfect for such a large object. Most telescopes don't offer a wide enough view to show the cluster in its "natural environment (Fig. 4.10)."

Messier 44 is the perfect binocular deep-sky object because it lies only 610 ly away. More than 20 of its stars shine brighter than mag 8. Its true diameter of 15 ly covers 1.2° in the sky. Messier 44 is believed to consist of roughly 200 stars, of which over 100 are variable ones. These have exhausted the hydrogen supply in their cores and are leaving the main sequence toward the realm of the giants. Its hottest main sequence stars are A6-stars. Therefore M44's age is estimated at about 600–800 million years, which is comparable to the age of another open cluster, the Hyades[6] in the constellation of Taurus. Both clusters, which seem to share other similarities as well, are thought to have originated from the same giant

---

[6] See the December objects.

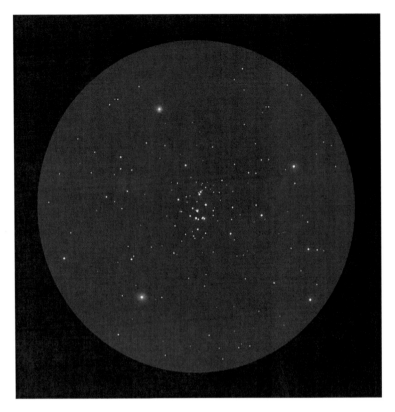

Fig. 4.10 The Beehive, observed with a pair of 8 × 56 binoculars. The fov is 5.5°

molecular cloud. If that is true, then the Beehive and the Hyades are the surviving remnants of an old and perished OB association. Although the Beehive is relatively nearby, stars with the luminosity of our Sun would still remain invisible in a pair of 50 mm binoculars. From within the Beehive, our Sun would shine with the feeble light of a mag 11 star. We can only see so far thanks to the highly luminous giants of our galaxy.

Now that you have observed the Beehive, try to see its shape with the naked eye. This is an excellent exercise to train your averted vision.

Messier 67

Cancer, though far from the galactic plane, harbors another interesting open cluster, called Messier 67. Our binoculars have no trouble locating this fine object. Just point your binoculars at α Cancri and move roughly 2° due west, as shown on the finder chart of Fig. 4.9. A pair of 8 × 56 binoculars show an elongated and mottled glow SW of a mag 8 star. Its shape measures 12′ and reminds one of a horn. The tip of the horn, which is located at the western end of the cluster, is brighter. A pair of 15 × 70 binoculars show a partial resolution. The cluster's brightest stars are of mag 9.9. They are located near the tip of the horn. A pair of 20 × 80 binoculars are capable of resolving even more cluster members. In such an aperture, the horn seems to feature wings as well. Some say the cluster, which now measures 25′ wide, resembles a winged sea horse (Fig. 4.11).

Unfortunately, the eighth magnitude star is not at all a member of M67. The cluster's lucida is a K3 giant in the tip of the horn. Messier 67 consists of over 500 stars. It has no bluer main sequence stars than those of type F. M67 is believed to be an extremely old cluster. Its age is estimated at 3.7 billion years, almost as old as our Solar System.

Messier 67 looks rather dim with an apparent magnitude of 6.9, compared to the Beehive, but that's only because it is 5 times further away. Messier 67 measures 21 ly across and lies 3,000 ly away. The cluster must lie roughly 1,500 ly off the galactic plane. This too is an exceptional measure for a galactic cluster. Some say that M67 has drifted this far off the galactic plane because it's very old, while others say that M67 has been protected from tidal destruction because of its remote location off the galactic plane.

A cluster as old as Messier 67 should have a reasonable number of red giants. Astronomers have indeed discovered many evolved red giants and around 150 white dwarfs. But despite the cluster's respectable age, a total of 11 hot B-stars have been found as well. Normally these blue stars are only present in very young clusters. In the case of M67, they are called blue stragglers. These stars are believed to have increased their masses due to interactions with other stars. One explanation is that a blue straggler could be formed by the merger of two binary stars. Their combined mass could create a single massive star. Another explanation is that mass transfer can occur in a binary system. The more massive companion will expand first to

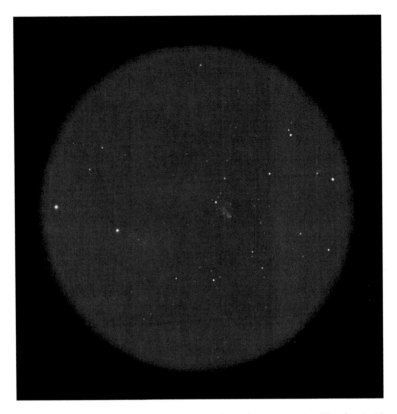

Fig. 4.11 Messier 67, observed with a 4 in. refractor at 16×. The fov is 3° wide. The bright star at the eastern edge of the fov is α Cancri

become a red giant. If that star grows large enough, its outer layers might overflow to the lighter companion, which will then become the more massive blue straggler.

Messier 67 also harbors 100 or so solar-type stars. These stars have the same chemical composition and mass and are almost the same age as our Sun. It appears that most of these stars are either significantly more active or much more quiescent than our day star. Messier 67 is therefore preeminently a target for the study of the behavior of stars much like our Sun.

MARCH

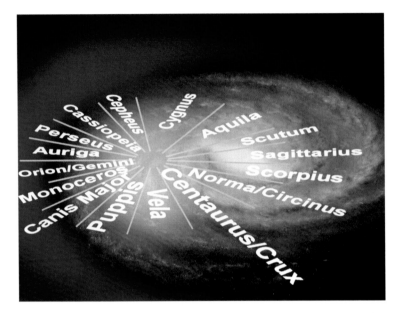

Fig. 5.1 During spring, the Sun is well below the galactic plane. Our night-time window is straight above the galactic plane and thus pointed towards the north galactic pole

March is a special month. This time of the year the Sun passes through the Vernal Point in the constellation of Pisces. It is an annual milestone that announces the beginning of the spring season on the northern hemisphere. The March equinox has been celebrated by several cultures as the New Year's Day of their calendar.

From a deep-sky observer's perspective, spring evenings offer a very deep window into intergalactic space. If you take a look at Fig. 2.3 you'll see around Earth the golden ring representing the ecliptic. The vernal point is

R. De Laet, *The Casual Sky Observer's Guide*, Astronomer's Pocket Field Guide,
DOI 10.1007/978-1-4614-0595-5_5, © Springer Science+Business Media, LLC 2012

on the opposite side of the autumnal point. Therefore, the Sun is well below the galactic plane, as shown in Fig. 5.1. Our night-time window to the universe during the following months is thus pointed in the direction of the North galactic pole. During spring evenings, the Milky Way and its dust-laden spiral arms are located near our local horizon.

Because this stretch of the sky above the galactic disk is relatively clear of interstellar matter, we can easily discover a new type of deep-sky object, the galaxies! These are the large aggregations of hundreds of billions of stars and interstellar matter, similar to our own Milky Way.

Observing galaxies with a pair of binoculars is like looking for seashells in the cosmic ocean. However, even with such modest equipment, galaxies prove to be social creatures, floating through the dark emptiness of space in close associations called clusters.

How about the Milky Way then, is it part of a cluster, too? The answer, which might surprise you, is – yes, the Milky Way is part of a cluster called the Local Group, which consists of over 30 galaxies.[1] After the Andromeda Galaxy (Messier 31), our Milky Way is the most massive galaxy of the remaining galaxies in the Local Group. Figure 5.2 shows the distribution of the featured galaxies near our Milky Way. The orientation of this sketch is identical with the orientation of the galaxy views. The blue grid of Fig. 5.2 is thus parallel with the galactic plane of Fig. 2.3. Therefore the constellation sectors of Fig. 2.4 apply here as well.

You can regard Fig. 2.3 as a close-up view of Fig. 5.2. Keep in mind that the galaxies of Fig. 5.2 are exaggerated in size and brightness to make them stand out clearly in the illustration. The three largest galaxies of the Local Group – M31, M33 and our home galaxy – are shown right above the blue plane. The Ursae Major galaxy group is represented by M81, M82 and NGC 2403. The constellation of Leo has for binocular observers two galaxy groups to offer, the M65/M66 Group and the M95/M96 group. The Coma Berenices/Virgo border is extremely rich in galaxies. Here we find the

---

[1] Most of the galaxies of the Local Group are dwarf galaxies.

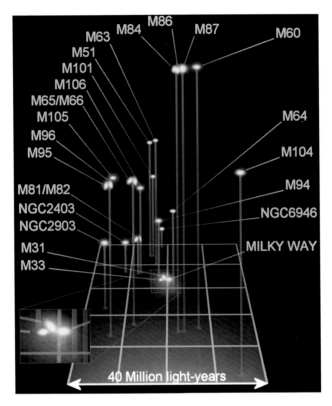

Fig. 5.2 The neighborhood of the Milky Way with all the featured galaxies. The *blue* plane is parallel with the galactic plane. Neither the size nor the brightness of the galaxies is proportional to their relative distance

Coma-Virgo galaxy cluster,[2] represented by M84, M86 and M87. All the featured galaxies are members of the so called Local Galaxy Supercluster. The Local Group, the one that the Milky Way belongs to, is considered to lie at the edge of the Local Galaxy Supercluster. The Virgo Cluster, which is approximately 59 million ly away, is believed to be the dense heart of the Local Galaxy Supercluster. It is exciting to realize that many members of the Local Galaxy Supercluster can be seen with a common pair of binoculars.

---

[2] Often called the Virgo Cluster.

Of course there are many more galaxies in the Milky Way's neighborhood, but only the ones featured in this book are shown in Fig. 5.2.

Observing galaxies with binoculars requires at best a moonless night, a dark sky and averted vision. Galaxies appear like pale and diffuse streaks of light in the eyepieces of your equipment. Nevertheless common binoculars are capable of revealing the presence and even the shape of many galaxies, as you will notice on the following tours. This month, we will visit the brightest galaxies of Ursae Major, Camelopardalis and Leo.

Messier 81 and Messier 82

It seems appropriate to start our galaxy hunt with a bright and easy target such as Messier 81 in Ursae Major. First we need to locate the mag 4.6 star 24 Ursae Majoris. Once again we'll use the Big Dipper as our starting point. This season, the Greater Bear is high overhead, as shown on the finder chart of Fig. 5.3. You can follow the dashed line from γ UMa over α UMa and move another 12° towards 24 UMa.

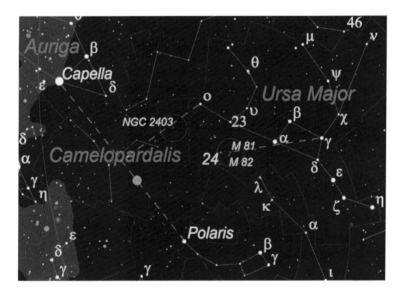

Fig. 5.3 Finder chart for M81/M82 and NGC 2403. The orientation of the constellations is appropriate for the spring season

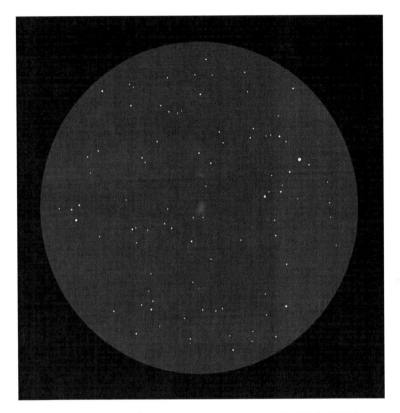

Fig. 5.4 The fine duo of galaxies M81/M82, observed with 8 × 56 binoculars. The bright star west of the duo is 24 UMa

With 24 UMa near the western edge of the fov of your binoculars, M81 should be visible in the middle of the field. Its companion, M82, can be glimpsed at about 40′ to the north of M81. Don't forget that when you face true north, Polaris is *under* your target this time of the year. Thus south is up and west is left in your eyepieces. The sketch of Fig. 5.4, however, has north up and west right, as usual.

The famous M81 and M82 pair of galaxies creates a stunning duo in any pair of binoculars. M81 is the brighter one of the two. It looks like a small glowing patch of light. Its center appears to be a tad brighter. Higher power instruments will show its stellar nucleus. M82 is rather weak, but the cigar shape is clearly present! Bear in mind that the stars in the field of view are all members of our own galaxy. M81 and 82 are to be imagined 10,000 times further away than the field stars you see. It seems amazing that from a distance of 12 million light-years, this duo of galaxies is still recognizable in a simple pair of binoculars. The light that these distant islands emit has been traveling for 12 million years through intergalactic space before it gets captured by our eyes on a crisp spring evening.

Messier 81 is the largest galaxy of the M81/M82 group, a group of over 30 galaxies. M81's apparent brightness of mag 6.8 is spread over an area of 20' × 10'. This grand design spiral galaxy has a true diameter of 92,000 ly and an integrated absolute brightness of mag −21.1 (compared to the value of mag −20.6 of the Milky Way).

Messier 82 is a real and very close companion of M81. It has an apparent brightness of mag 8.4 and measures 9' × 4'. M82, which is 37,000 ly in size, is called a starburst galaxy, a galaxy in which star birth occurs at an extremely high rate. The reason for this is the close encounter of M82 with M81 about 600 million years ago. The tidal forces that came with this interaction have totally deformed Messier 82 and triggered the high rate of star formation. We see this galaxy as a spindle because of its nearly edge-on orientation. Much of the galaxy's activity is absorbed by thick clouds of dust. Although smaller than the Milky Way, Messier 82 is in reality five times as bright. M81 and M82 are separated by only 130,000 ly.

NGC 2403

NGC 2403 in Camelopardalis is one of Messier's missed objects. This galaxy is so bright that Charles Messier should have been able to pick it up in his instruments. The late Walter Scott Houston called it 'the brightest galaxy north of the celestial equator that does not have a Messier number.' That statement alone is a good reason to go after this galaxy with your small pair of binoculars. It may sound strange, but NGC 2403 is a member of the

M81 group of galaxies. Its distance to the M81/M82 duo is about three million ly, comparable to the distance between our Milky Way and the Andromeda Galaxy, M31.[3] With a mag of 8.4, NGC 2403 is almost as bright as M82. Its distance to our galaxy is about ten million ly. But NGC 2403 appears much larger (17′×10′) and has therefore a lower surface brightness (14.6 compared to 12.8 for M82). The galaxy's disk is inclined at about 30° from edge-on.

NGC 2403 can be found from different starting points. Here is a popular approach for low power instruments and binoculars, as shown with the dashed lines on Fig. 5.3: draw an imaginary line from Capella to Polaris. Both stars are roughly 45° apart. Then find the midpoint of that line. From that point, draw a line (the bisector) perpendicular to the first line, for about 15° in the direction of the Bear's nose (Omicron Ursae Majoris) until you run into NGC 2403. The galaxy lies seven and a half degrees northwest of Omicron UMa.

Some observers report that this galaxy is easily mistaken for a comet. Others have a very different impression. When first locating NGC 2403 with a pair of 8×56 binoculars, it may not remind you of a comet at all. NGC 2403 shows a fairly large and diffuse core, centered in a faint but obviously elongated halo. This appearance may remind you of NGC 2903 in Leo with a 4 in 'scope. It's definitely one of the more rewarding galaxies for binocular observers to locate! As with most of these objects, the more time you spend examining these distant islands, the clearer you will see them. The brightest star east of NGC 2403 is 51 Camelopardalis.

Be aware that when you observe NGC 2403 while facing true north, Polaris and north are located below your fov. This is an upside down orientation when compared to the sketch in Fig. 5.5.

NGC 2403 and M81 are both fine spiral galaxies. An observer from an inhabited planet from within one of these galaxies would actually have a similar view of our Galaxy and the Andromeda Galaxy.

---

[3] See the chapter on November objects.

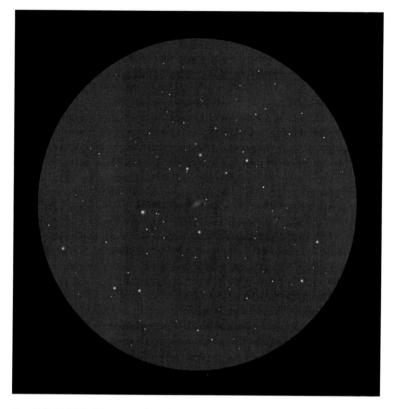

Fig. 5.5 NGC 2403, west of sixth mag 51 Camelopardalis, observed with a pair of 8 × 56 binoculars

Regulus

Our next targets can all be found in the constellation of Leo, the Lion. Use Fig. 1.3 to locate Leo. Here, too, you can use the Big Dipper as a signpost. If you draw a line from the two easternmost stars of the Dipper's bowl to the southwest for about 50°, you arrive at Regulus, the brightest star of the constellation of Leo. Regulus means 'the Little King,' a name that originated ages ago in ancient Mesopotamia. Our ancestors must have had high expectations of this majestic representation of the king of beasts, the lion. It also

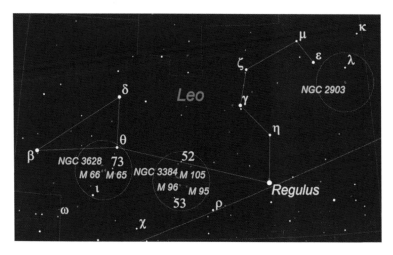

Fig. 5.6 Finder chart for the Leo deep-sky objects

did not go unnoticed to them that Regulus, the heart of the Lion, was strategically placed near the ecliptic. Of the brightest stars in the sky, Regulus is closest to the ecliptic and is regularly visited by planets and sometimes occulted by the Moon. Such events are always a joy to observe (Fig. 5.6).

Regulus, or Alpha Leonis, is an interesting object for the binocular astronomer. Regulus forms a wide magnitude-contrast duo with a much fainter companion. Their separation is a fair 3 arc minutes. But it's a challenging task to discover the 8th magnitude companion in the glare of the 1.3 magnitude Regulus. Bear in mind that Regulus is 400 times brighter than its companion. Regulus is a blue-white main sequence B7-star with an absolute brightness of mag −0.5, which is 150 times brighter than our Sun. It has 3.5 times the mass of our Sun. B-stars are relative rare stars. Only 1 in 800 main sequence stars in the Sun's environment is a B-star. Regulus appears to be a fast spinner, with a rotation period of 16 h. Because of this high rotation speed, the star has an oblate form. If spun any faster, it would literally fly apart.

Astronomers have discovered another companion very close to Regulus. This small companion is believed to be a white dwarf. The theory goes that the companion once was a red giant that transferred most of its mass to

Regulus. This could explain Regulus' abnormal rotation speed. The eighth magnitude companion, visible in binoculars, appears to be a double system, too. This makes Regulus a quadruple star system. With a distance of only 77 ly, Regulus is the closest B-star in the Sun's neighborhood (see also Fig. 4.2). Our day star, if it was placed next to Regulus, would shine with the brightness of a mag 6.7 star.

The brighter 'star' at the left edge of the sketch in Fig. 5.7 is the mighty planet Saturn, which happened to visit Regulus at the time of this observation.

Fig. 5.7 A binocular observation of Regulus, the heart of the Lion, with a pair of 8 × 56 binoculars. Regulus' fainter close companion is visible as well. The bright object at the left edge of the fov is the planet Saturn

It is most likely that Saturn will not be around when you observe Regulus. Planets are too small to show any detail in binoculars. That's why you can't see Saturn's rings, either. However there are occasions when Saturn's moon Titan can be detected with a pair of binoculars, but at the time of the observation, Titan was too close to Saturn to be discerned. A conjunction between Regulus and a planet always shows a marvelous color contrast.

NGC 2903

The constellation of Leo is famous for its spur of distant galaxies. One of the brighter samples is NGC 2903, a fine object for binoculars and small telescopes. It would have been a worthy member of Messier's catalog, compared with the other entries from the constellation of Leo. This solitaire spiral galaxy at mag 9 lies 30 million ly away. It measures 11′ × 5′. NGC 2903 is comparable to our own Milky Way. It has about the same size, and it also features a central bar. However, the core of this galaxy is extraordinarily bright. NGC 2903 is called a hot-spot galaxy. Investigations with the Hubble Space Telescope have revealed a series of hot spots in the core of this galaxy. These hot spots are peculiar blue clusters of young and bright stars. It seems that the gravity of NGC 2903's central bar has accelerated the star formation in the galaxy's core.

Locating the hot-spot galaxy is rather simple: it can be found at 1.5° south of fourth magnitude Lambda Leonis, near the nose of the Lion. The galaxy's core and stellar nucleus are the first features that show up. A pair of 8 × 56 binoculars reveal a fuzzy star only with averted vision. With 10 × 50 binoculars, the core appears elongated in a north–south direction. The galaxy might disappoint you at first and look rather small. A dark sky will help to see the galaxy better. But with patience and keen averted vision, the faint but large halo starts to light up against the background glow. The galaxy seems to grow in the eyepiece, though, when you take the time to make a sketch. A 4 in. refractor shows the elliptical shape of NGC 2903 clearly. The blazing core of this hot-spot galaxy stands out. There are hints of the central bar, too. These subtle features might remain invisible for novice observers. The more experience you gain behind the eyepiece, the better you will see the true faces of these distant islands (Fig. 5.8).

Fig. 5.8 NGC 2903, observed with a 4-in refractor and a magnification of 63×. The fov is 65′

The M66 Galaxy Group

When the constellation of the Lion culminates, you should shift your gaze to the M66 group of galaxies. Although these galaxies are fairly easy to see with a small telescope, they offer a challenge for small binoculars. Aiming your binoculars is very easy, as Theta and Iota Leonis border the northern and southern edge of the binocular field of view. Under a mag 5 sky, you may only detect M66, which has a total magnitude of 8.9. M65, with a

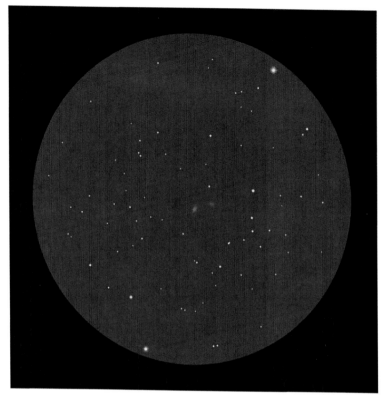

Fig. 5.9 The Leo duo of galaxies, M66 and M65, observed with a pair of 8 × 56 binoculars under a dark sky

magnitude of 9.3, remains invisible. The situation improves when observing from a darker site with a nelm of 6.0. From such a site, you may notice the oval glow of M66 immediately near the center of the fov. And with averted vision, you can spot the faint glow of M65. It always is thrill to realize that the light produced by this duo of galaxies traveled a distance of 36 million light-years to reach your eye (Fig. 5.9).

These interacting galaxies both measure roughly 90,000 ly across. Their true separation is about 200,000 ly. M66 and M65 are rather luminous, despite the fact that we see them from nearly edge-on. The M66 group has a third large member, NGC 3638, which is too faint to be seen with most binoculars. The group is often called the Leo Triplet of Galaxies. The Triplet can be observed with 15×70 binoculars under dark skies.

The M96 Group of Galaxies

Our final destination of this month's tour is a real challenge for your binoculars. Nonetheless their location is easy to find. As you can see on the finder chart of Fig. 5.6, this galaxy group is located right in the middle between Regulus and Iota Leonis. There you find the fifth magnitude 52 Leonis and 53 Leonis, respectively north and south of our target. Both stars will fit nicely in the fov of most binoculars. The visibility of the M96 group of galaxies, however, will depend on the quality of your sky and the type of binoculars you use. Only with a pair of 16×80 binoculars might you be able to observe the four interacting galaxies of this group, as shown on the sketch of Fig. 5.10. The brightest galaxy is the mag 9.2 Messier 96. It is the brightest spot on the sketch and measures roughly 6′ across. M96 is the only galaxy in the group that shows a noticeable bright core. The next brightest galaxy is Messier 105 with a magnitude of 9.3. M105 is the hazy little dot 48′ north of M96. Its apparent size is only 2′. One can expect that Messier 95 would show up next, but that's not the case.

NGC 3384 is definitely number three on the visual brightness scale, although its integrated brightness of mag 9.9 is inferior to the one of M95. Perhaps both M105 and NGC 3384 are easier to see because of their intense glowing cores. They form a close physical pair, as NGC 3384 lies only 7′ east of M105. Messier 95 is the trickiest galaxy to hunt down. Its integrated brightness of magnitude 9.7 is spread over an area four times as large as NGC 3384. Therefore M95's ethereal glow, 41′ west of M96, can only be detected with the greatest care. Messier 95, which shows no core at all, seems to grow larger than M96. It is a difficult task to notice the four galaxies together in one glance, as M95 is often lost in the background glow.

The M96 group of galaxies lies about 35 million ly away. M96 and M95 are spiral galaxies, like our Milky Way. They measure 75,000 ly across. M105 on the other hand is an elliptical system with a size of 55,000 ly. M95 and M105

Fig. 5.10 The M96 group of galaxies is a rewarding target for large binoculars. The sketch was made after viewing with a pair of 16×80 binoculars. The fov measures 3° across

are both roughly 500,000 ly away from M96. If M105 and NGC 3384 are equally distant from us, then their true separation is only 70,000 ly.

The galaxies of Leo can be used as a yardstick for measuring the quality of your sky during the months of March, April and May. The visibility of Leo's galaxies will depend on the sky's darkness and transparency. Looking up at these galaxies before moving on to other deep-sky objects will assist you with rating the sky of your observing site.

All the galaxies featured on this March tour, as well as our own Milky Way, are members of the Local Galaxy Supercluster. We will visit the gripping center of the Local Galaxy Supercluster, located an overwhelming 55 million ly from us in the direction of the Virgo/Coma border, during the next month.

CHAPTER 6

APRIL

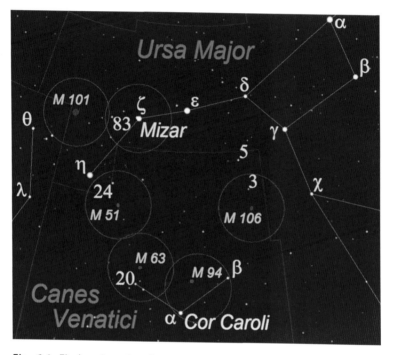

Fig. 6.1 Finder chart for the deep-sky objects around Ursa Major and Canes Venatici

This month we continue with our voyage through intergalactic space in the direction of the north galactic pole. Our night-time window is illustrated by Fig. 5.1 of the previous chapter. The constellations of interest are Ursae Major again, Canes Venatici, Coma Berenices and Virgo. These constellations are strewn with lots of galaxies, which are all part of the Local Galaxy Supercluster. Figure 5.2 illustrates the featured galaxies in perspective with

R. De Laet, *The Casual Sky Observer's Guide*, Astronomer's Pocket Field Guide,
DOI 10.1007/978-1-4614-0595-5_6, © Springer Science+Business Media, LLC 2012

our Milky Way. But this month we will not only encounter faraway galaxies. On one of our trips we will visit a remarkable nearby galactic cluster as well, Melotte 111.

Cor Caroli and Messier 94

The signpost for this month's first tour is the brightest star of the constellation of Canes Venatici, the Latin name for the Hunting Dogs. This small constellation is located south of the handle of the Big Dipper, as is shown in Fig. 6.1. α Canum Venaticorum is known as Cor Caroli, which means 'Charles' Heart.' The name was given in honor of King Charles II of England. Cor Caroli is a binary star at a distance of 110 ly (see also Fig. 4.2). Its primary is a mag 2.9 A0 main sequence star, famous for its strong magnetic field and its weird chemical composition. The secondary star of mag 5.9 can only be seen with a telescope. Both α CVn and β CVn fit in the low power fov of ordinary binoculars.

These two stars show us the way to Messier 94, a bright mag 8.2 spiral galaxy at a distance of 16 million ly. Figure 6.2 shows that M94 appears like an out-of-focus star, roughly 3° NNW of Cor Caroli. Locating M94 is not that hard. First aim your binoculars right at the middle between Cor Caroli and β CVn and then move one and a half degrees to the NE. Messier 94 should be visible now in the middle of your fov. A pair of 8×56 binoculars show a 5' wide out-of-focus star. We see Messier 94 face on. This spiral galaxy is only half as large as our Milky Way and yet it shines with an absolute brightness of mag −20.5. The reason for the exceptional brightness is that Messier 94 is a starburst galaxy.

Astronomers have discovered two concentric rings of high star formation activity in this galaxy. This makes M94 a rare galaxy. The reason for the galaxy's high activity is still unclear. The galaxy's peculiar appearance can be observed with a small telescope. A 4-in. refractor shows a non-stellar nucleus in a well defined ring-shaped core. The darkest of nights allow us to see an extended tenuous halo as well. Ordinary binoculars show that M94 suddenly brightens towards its center. Observing galaxies with binoculars is an excellent way of training your vision. Galaxies may all appear as small smudges of light at first look, but they never really look the same. Examine the sketches of the galaxies in this book carefully and you'll see that each one has a different look.

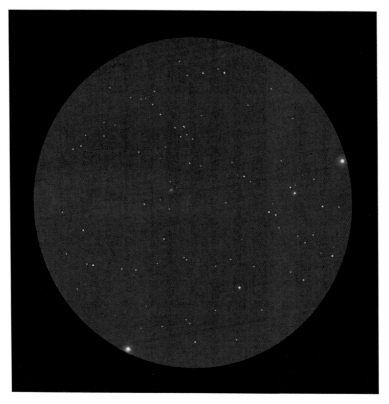

Fig. 6.2 Cor Caroli, β CVn and M94, observed with a pair of 8×56 binoculars

With both the bright Cor Caroli and the fuzzy star-like image of M94 in your binoculars' sight, try to imagine that the fuzzy star lies one and half million times further away than Cor Caroli.

Messier 106

Our next target, Messier 106, is a very rewarding object in any kind of instrument. This grand spiral galaxy shines with an apparent brightness of mag 8.3 and it measures 11′×5′. Unfortunately, there are no bright stars nearby to use as a signpost.

Fig. 6.3 Messier 106 observed with a 4-in. refractor and a magnification of 63×. The fov is 65′ in size

Binocular observers can use the large fov of their instrument to their advantage in locating M106. The simplest way to locate M106 is to move in a straight line from β CVn to γ UMa (see Fig. 6.1). Messier 106 will show up brightly halfway in between, at 1.5° south of the fifth magnitude star 3CVn. This galaxy has something to offer in any instrument. A pair of 8×56 binoculars show M106's bright core sharply. With averted vision, the galaxy's slim halo grows to a 7′ long spindle of light. The galaxy looks very fragile and begs for a more patient study.

With growing aperture, more subtle details become apparent. The view with the 4-in. refractor is very complex. At low power, the galaxy looks elongated in a N–S direction. At first sight, the nucleus appears stellar. But then, with a closer look, the nuclear region seems to be broken in pieces. The northern spiral arm is the most conspicuous one. It can be followed from the core to the northern edge of the halo. There are some traces of the southern arm visible as well, but they do not stand out that easily against the mottled glow of the galaxy's halo (Fig. 6.3).

An interesting exercise is to look for dark features or patches within the soft glow of the halo. Don't expect these details to pop into view right away. It is only with a patient examination of the galaxy's appearance with your averted vision that you can confirm such subtle features. Messier 106 lies 26 million ly away and measures 130,000 ly across. It has a slightly higher absolute brightness than our galaxy. It is one of the most beautiful galaxies for small instrument observers. Only Messier 31 exceeds it in beauty, but that galaxy lies ten times closer to us, too.

Melotte 111

During spring evenings we look straight up from within the galactic plane in the direction of the north galactic pole, in the constellation of Coma Berenices. Therefore we look through the most transparent region of the galactic disk into the vast intergalactic space. Here there are no dense galactic spiral arms or giant molecular clouds that block the view from beyond. That's why we can see so many distant galaxies during this time of the year. And yet, the highlight of this month is… a galactic cluster, called Melotte[1] 111. In fact Mel 111 is a rather large and bright cluster, too, called the Coma Star Cluster because it is located in the constellation of Coma Berenices. The constellation represents the locks of the Egyptian queen Berenice, sacrificed to the goddess Aphrodite for the safe return of her husband from war. Now, there aren't many clusters that have a whole constellation named after them, so be prepared to admire one of the most attractive binocular clusters in the skies. Use Fig. 6.4 to locate Coma Berenices, which is flanked by bright Arcturus in the east and the hindquarters of Leo in the west.

---

[1] Philibert Jacques Melotte was a British astronomer who published a catalog of star clusters in 1915.

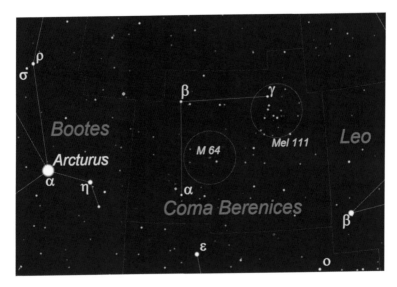

Fig. 6.4 Finder chart for the coma objects

Mel 111 has an apparent brightness of mag 1.8 and measures a full 5° across. The Coma Star Cluster ironically highlights the direction of the north galactic pole. Mel 111 seems to be completely thrown out of the Milky Way, but that's only an illusion. The cluster hovers only about 284 ly above the Sun and is definitely part of the smooth galactic disk of stars. Mel 111 spans 23 ly and contains roughly 260 stars. Many of its stars can be seen with the naked eye from a dark location.

Bear in mind that the Coma stars you see are much brighter than our Sun. From within the Coma Star Cluster, our Sun would have an apparent brightness of mag 9.5. The cluster is known for its lack of red dwarfs, which are thought to have 'evaporated.' Mel 111 does not exhibit enough gravitational attraction to bind its members. Therefore the low-mass stars can easily escape into the smooth galactic disk of stars. This lightweight cluster, which is some 600 million years old, will eventually completely disintegrate into the galactic plane. Wide-field binoculars are the ideal instruments to admire this delightful cluster. A pair of 8×40 or 8×56 binoculars show Mel 111 at its best. The magnification of 15×70 binoculars disperses the cluster into a very loose gathering of stars. Try also to find the few double stars that Mel 111 has to offer (Fig. 6.5).

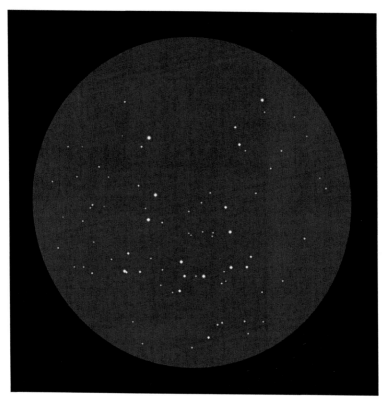

Fig. 6.5 Melotte 111, south of γ Comae, observed with a pair of 8×56 binoculars

Messier 64

From Melotte 111, it is only a small star hop to M64 (the Black Eye Galaxy). This remarkable spiral galaxy lies on a straight line drawn from α Comae to γ Comae. Our starting point is α Comae. From here move one binocular field or 5° towards γ Comae. A fifth magnitude star, named 35 Comae, should appear in your view. M64 lies roughly 1° ENE of 35 Comae.

A pair of 8×56 binoculars show this mag 8.5 galaxy as an elongated patch, measuring 6'×3'. M64's brightness increases slightly towards its nucleus.

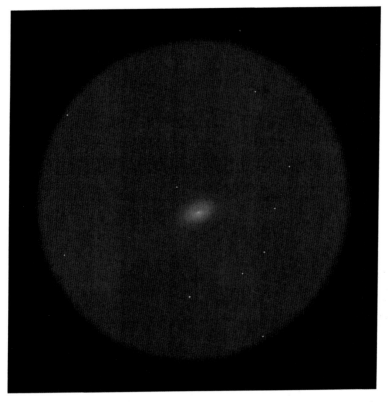

Fig. 6.6 Messier 64, observed with a 4-in. refractor at 100×. The fov is 40′

The galaxy reveals some of its peculiar characteristics with larger instruments, as shown on the sketch in Fig. 6.6. In a 4-in. refractor, the galaxy's bright nucleus appears elongated, too. M64's core is well defined also. The galaxy's dim halo is difficult to measure, because it smoothly dissolves into the background.

The 'highlight' of Messier 64 is its famous 'dark eye.' It's a rather tricky feature to discern visually in a small telescope, although photographs show it prominently. If you take the time to study the region around the nucleus with higher magnification, it should become clear that the area north of the galaxy's nucleus appears a tad darker than its surroundings. That's the Black Eye. Taking notes and making a sketch will help to draw your attention to such inconspicuous features. M64 is an object whose appearance depends on the magnification used. Its delicate halo is best noticed with low power, while the Black Eye demands a higher magnification.

Messier 64 is about 56,000 ly large and lies 18 million ly away, which is a third of the distance to the center of the Local Galaxy Supercluster (see Fig. 5.2). The Black Eye is a complicated cloud of dark matter, roughly 5,000 ly in size, that blocks our view of the galaxy's disk. The theory goes that it is the remnant of a satellite galaxy that has been cannibalized by M64. Evidence for this theory might be found in the fact that the stars in the inner 3,000 ly disk of M64 rotate in the opposite direction of the stars outside of this region.

The Virgo Cluster of Galaxies

Our next destination lies far, far away, at an incredible distance of 55 million ly in the constellation of Virgo. Here we find the most remote galaxies that can be observed with ordinary binoculars: the Virgo Cluster of galaxies. The light that our binoculars capture from these galaxies departed long before humans populated Earth.

Although our trip goes back into time, but we should not consider the Virgo Cluster to be at a remote location. On the contrary, it is at the very core of the Local Galaxy Supercluster (LSC). The LSC is believed to have a diameter of 110 million ly. A total of 100 or so galaxy clusters are member of the LSC. One of them is our own Local Group, at the very outskirts of the LSC. The galaxies shown in Fig. 5.2 represent only a small section of the LSC.

Let's have a look at the core of our 'local neighborhood,' the Virgo Galaxy Cluster. The Virgo cluster is believed to contain between 1,300 and 2,000 galaxies. From our point of view, the Virgo cluster spans about 8° of sky near the border of Virgo and Coma Berenices. Its true width is estimated at roughly eight million ly (Fig. 6.7).

Locating the Virgo cluster is fairly simple, as it lies on a straight line from Vindemiatrix (ε Virginis) towards Denebola (β Leonis). When you start from Vindemiatrix, use the fifth magnitude ρ Virginis as a supplementary signpost.

Fig. 6.7 Finder chart for the Virgo cluster of galaxies

From Rho Vir, it's a 1° 21′ jump to the north to find Messier 60. This lenticular galaxy looks like a small fuzzy disk measuring 2′ across. Nevertheless, with its small appearance and apparent brightness of mag 8.8, M60 is a giant elliptical system with a true diameter of 115,000 ly and an absolute magnitude of −22.

If you move your binoculars 3° to the west, another galaxy comes into view – the dominant supergalaxy Messier 87. This mag 8.6 elliptical galaxy spans 132,000 ly. You might consider that our Milky Way is not a great deal smaller, but ellipticals have much more volume, because spiral galaxies are really just flattened disks. M87 contains 200 times the mass of our galaxy and is the most massive galaxy ever discovered. Its integrated absolute brightness is −22.4. The galaxy is famous for its supermassive black hole, hiding at the center of M87. Pictures taken with large telescopes have revealed that this black hole ejects matter into deep space at high speeds in a concentrated jet. M87 is believed to contain over 15,000 globular clusters. Theory goes that these globulars are the remaining nuclei of 'swallowed up' galaxies. Given M87's mass and size, the voracious galaxy might frequently have cannibalized other galaxies, whose stars were absorbed by the spherical halo of giant M87. From 55 million ly away, this central Virgo galaxy is clearly visible in ordinary binoculars. A pair of 8 × 56 binoculars show a fuzzy blob of light, like an out-of-focus star. With patience, M87's condensed nucleus can be detected as well. An eighth mag star lies 6' to the north of Messier 87.

Now move your fov another degree to the west. Can you see 2 minute, luminous spots, 20' apart? This duo of elliptical galaxies are called Messier 86 and Messier 84. Both objects are of ninth magnitude. M86 has a diameter of 147,000 ly while M84 measures 110,000 ly across. These, too, are giant galaxies, but somewhat less massive than M87. The true brightness of these outrageous Virgo galaxies is beyond measure. But we can try to compare these galaxies with the one we are most familiar with, the Milky Way. Alas, our mighty galaxy proves to be a dwarf, compared with, for example, M87. The apparent brightness of the Milky Way would be less than mag 10.5 if placed next to M87.

Ordinary binoculars are capable of showing these four extraordinary cosmic powerhouses in a single fov (see Fig. 6.8). They are supermassive, insatiable galaxies, representing the heart of the Virgo cluster at the center of the Local Galaxy SuperCluster, 55 million ly away. From this incredible distance, the massive core of the LSC exhibits its strong gravitational force on our Milky Way, like on every other galaxy within its range.

The LSC is only one of millions of superclusters in the observable universe. Superclusters are distributed in space like filaments in a foam-like structure, often called the Cosmic Web. Superclusters are separated by vast voids, comparable to the large cavities found in a sponge. The Cosmic Web is today the largest known structure in the universe. On the grand scale of the Cosmic Web, the daily concerns of us humans seem rather insignificant, don't they? But then again, it's a testimony to the reasoning power of the human species that we can wonder about the mystery of our fragile existence in the compelling immensity of the cosmic ocean.

Fig. 6.8 The core of the Virgo cluster of galaxies with M60, M87, M86 and M84. The bright star near the lower left edge is ρ Virginis. Observed using a pair of 8 × 56 binoculars

Messier 104

We end this month's tour with a visit to the brightest galaxy in the constellation Virgo, the famous Sombrero Galaxy (M104). This eighth mag spiral galaxy appears merely bright because it lies in the foreground of the Virgo cluster, at a distance of 35 million ly. Its diameter is estimated at 105,000 ly. Photographs show a very distinct ring of dust around the galaxy's central

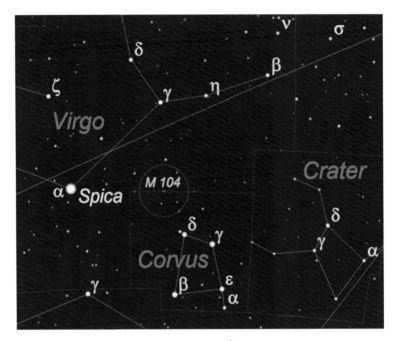

Fig. 6.9 Finder chart for M104

bulge. We see the galaxy's disk edge-on. The dust lane appears to cut through the galaxy's core, hence the look of a Mexican sombrero. Large telescopes are capable of displaying the 'sombrero look' visually. Small telescopes only show a hint of the dust lane, while binoculars merely reveal the galaxy's elongated orientation.

M104 is not an easy object to locate. It lies in a desolate region near the border of Virgo with Corvus. So have a good look at the brighter stars of Corvus to get acquainted with that part of the sky. With Fig. 6.9, you can imagine a straight line from δ Corvi to γ Virginis. M104 is only 5°30′ separated from δ Corvi. Take into account that this object never rises high above the southern horizon. This means that you will have to look through a thicker layer of air and that your window of opportunity is rather small. So be prepared to choose the crispest of nights to hunt down M104. The galaxy is visible with a pair of binoculars, but they'll only show the galaxy's

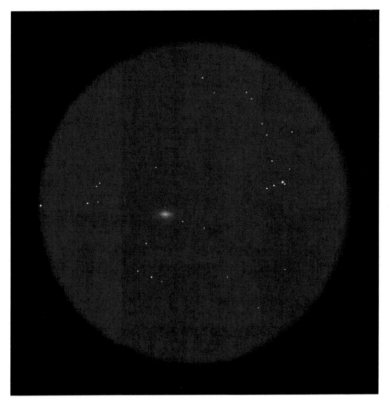

Fig. 6.10 Messier 104, observed with a 4-in. refractor at 63×. The fov is 65′ wide

bright central bulge as a minute fuzzy star. A small telescope reveals more details, as shown in Fig. 6.10. The patient observer will notice the east-west orientation of the galaxy's disk. You might also notice a 12th mag star on top of the galaxy's bulge. The galaxy's halo measures only 4′ in the sketch (see Fig. 6.10).

Messier 104 might not impress you at first, but it really is a bright galaxy. Our Milky Way would shine with an apparent magnitude of 9.6, compared to M104 eighth magnitude. The reason for M104's brightness could be found in the fact that a quarter of the galaxy's total mass is concentrated in its central bulge, while our Milky Way only has a seventh of its mass gathered centrally. Telescopic observers are rewarded with a small bonus. M104 is accompanied by a fine asterism, called Jaws. The asterism is shown near the NW edge of the sketch. Its brighter stars represent the teeth of the shark. It seems that Jaws is making a turn towards M104.

MAY

Fig. 7.1 Mizar, Alcor and Sidus Ludoviciana, observed with a 8×40 binocular. The fov is 8° wide

R. De Laet, *The Casual Sky Observer's Guide*, Astronomer's Pocket Field Guide, DOI 10.1007/978-1-4614-0595-5_7, © Springer Science+Business Media, LLC 2012

Our night-time window is slowly turning away from the north galactic pole towards our own galaxy. Figure 5.1 illustrates the galactic plane, with the Sun well below it during spring. Late night observers can already witness the rising of the summer constellations and the Milky Way with them, in the east. This month we'll take the opportunity to pay a last visit to intergalactic space, before the vast molecular dust clouds of our home galaxy move into view again. For a better understanding of the positions of the described galaxies, Fig. 5.2 illustrates the immediate neighborhood of the Milky Way.

Collinder 285

Our first object of this month's tour is a very famous one, the Big Dipper itself. Those who are not familiar with the Big Dipper can examine the northern sky charts in Figs. 1.4, 1.6, 1.8 and 1.10. The Big Dipper played a major role in the history of the constellations. The oldest traces go back to 3,000 B.C. in Sumeria, where the Big Dipper was the representation of a celestial chariot.

The Big Dipper (aka Collinder 285) is not just a chance alignment of bright stars, as five of the seven Dipper stars share the same proper motion in space. They are ζ, ε, δ, γ and β UMa (see Fig. 6.1). And if objects move in the same way, astronomers believe that they most probably share the same origin. If that is true, then Cr 285 is within a distance of 80 ly, the closest 'cluster' to our Sun. About 14 cluster members make up the 'core' of this rather sparse aggregation of stars. Our Sun is not a member of this system. We just happen to drift through the outer reaches of Cr 285, which is about 500 million years old.

The five Dipper stars of Cr 285 are all very bright A-stars, which are clearly visible with the naked eye on every clear night. Their true brightness is recognized when we compare these celestial lighthouses with our day star. The Sun would shine at a meager mag 6.7, when placed near these brilliant white Dipper stars. Therefore it would remain invisible to the naked eye. If you have trouble seeing the Dipper as a cluster, compare this sparse cluster with M45, the Pleiades (see Chapter 14). They, too, would appear very similar in our sky when placed six times closer to us. The true location of Cr 285 is represented in Fig. 4.2 by Mizar and Alcor, also known as ζ UMa. Mizar and Alcor will be described in greater detail below.

Mizar, Alcor and Sidus Ludoviciana

The Big Dipper harbors the legendary double star ζ UMa, which is the middle star of the Dipper's handle (see Fig. 6.1). Both companions have a proper name. The second mag primary is called Mizar, the fourth mag secondary is called Alcor. Several ancient cultures used Mizar and Alcor as a test of good naked-eye vision as they are only 12′ apart. In modern times, most people have no problem with splitting the famous double without a telescope. Mizar and Alcor are a splendid sight in even the smallest binoculars. They both belong to Cr 285, the Ursa Major Moving Cluster. The position of Mizar and Alcor with respect to the galactic plane is illustrated by Fig. 4.2.

Mizar itself is also part of the first known binary system. Mizar A and B are clearly split in a small telescope at 40×. But that's not the end of the story. Both Mizar A and B are doubles, too. This makes Mizar a quadruple star. Recently, astronomers have discovered that Alcor has a companion as well. Thus Mizar and Alcor are at least a sextuplet system (Fig. 7.1).

Binocular observers will notice a third star between Mizar and Alcor. This eighth mag star was noticed by the German professor Johann Liebknecht in 1722. Liebknecht was convinced that he had discovered a new planet. In an attempt to flatter his sovereign, Ludwig of Hessen-Darmstadt, Liebknecht called his newly found 'planet' Sidus Ludoviciana. Of course, Liebknecht was mistaken about it being a planet. But Sidus Ludoviciana offers a nice comparison with our Sun, which would shine a full magnitude brighter if placed between Mizar and Alcor.

Messier 51

From Cr 285, we'll leave our galactic backyard and make a giant leap into intergalactic space. Our next destination is a pair of interacting galaxies in Canes Venatici, called Messier 51. Locating M51 is easier than it seems. If you take a close look at Fig. 6.1, you'll find a straightforward method for locating M51. Our starting point is Mizar. From there we star hop 7° to the star at the tip of the Bear's tail, η UMa. Now make a 90° turn in the direction of Cor Caroli and note the angular distances. M51 is only 4° separated from η, which means that both will fit in the same binocular field.

Now don't expect to see a giant spiral blazing in your eyepieces, even though M51 is called the Whirlpool. This mag 8.4 pair of galaxies only measures 7′ across in your binoculars. The largest of the two is called NGC 5194, which we are fortunate to see face on, although from 27 million ly away. Therefore we have a clear view of its spiral structure.

NGC 5194 is a smaller version of our home galaxy. It is 87.000 ly in size but has only 15% of the mass of our Milky Way. The other, smaller galaxy, which is called NGC 5195, measures 43.000 ly across. NGC 5195 is believed to lie behind one of the spiral arms of the main galaxy. The close encounter of both systems probably triggered a high rate of star formation, especially in the smaller galaxy.

M51 is well known for its beautiful spiral arms. Unfortunately, it takes a large telescope to see all its splendor visually. Figure 7.2 shows what you can expect to see with a small telescope. Both galaxies show a distinct nucleus within a fainter halo. It will take some time to discern any detail at all in the halo of the main galaxy. You can confirm the brighter arcs within the large halo of M51 on a dark night. Do pay attention to the space between both galaxies, and see if the halos touch or not. Making a sketch of M51 will help you to observe it more precisely. Our Milky Way, when placed next to M51, would shine weaker at mag 9. The starburst, due to the interaction of both companions, might be responsible for M51's higher level of brightness.

M51 does not travel alone in space. It is the brightest member of the M51 group of galaxies, of which M63 is also a member. The other members are too faint to observe visually. Figure 5.2 shows how close in space M51 and M63 are.

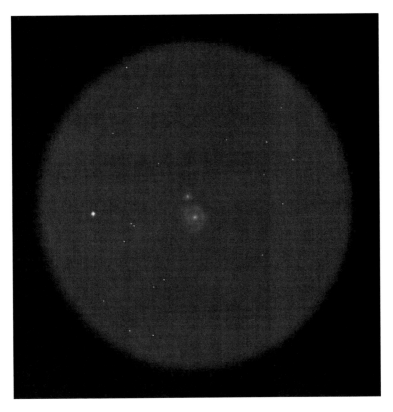

Fig. 7.2 Messier 51, observed with a 4-in. refractor and a magnification of 63×. The fov is a large 65′

Messier 63

M63, the Sunflower Galaxy, is the second Messier object that belongs to the M51 group of galaxies. This mag 8.6 binocular object is in fact a relatively close companion of M51. There is only 5° of an arc separating them in our sky. Locating M63 is easy with the use of Fig. 6.1. You can star hop over 5° from Cor Caroli to the fifth mag star 20 CVn. With 20 CVn centered in the eyepieces of your binoculars, look for a fuzzy, elongated dot at 1.5° north of 20 CVn. With care, the galaxy's core can be discerned as the bright center

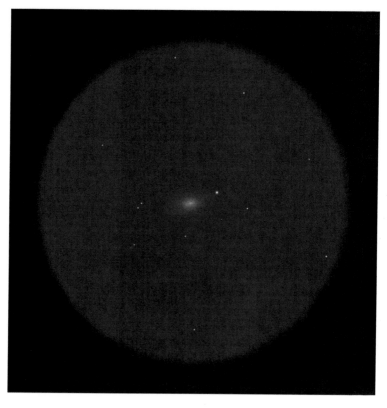

Fig. 7.3 M63, observed with a 4-in. refractor and a magnification of 100×. The fov is a large 41'

of the fuzzy glow. Its 6' large halo can be traced with binoculars as well. Its orientation is east–west. Note the mag 8.5 field star at the NW edge of the galaxy. A 4-in. refractor should reveal M63's features in greater depth, as is shown in Fig. 7.3. You can easily detect M63 at lowest power of 17×, next to the bright mag 8.5 field star. With each higher power, the view becomes more interesting but also more confusing.

M63 displays a tenuous and elongated halo that gradually brightens towards the core. The nucleus appears much brighter again and is definitely not stellar. It looks rather granular. The more time you spend at the eyepiece, the better you can see the distinct mottling in the faint halo. It appears that the halo grows larger with time. The mottling in the halo suggests that the core is surrounded by concentric oval rings.

M63's appearance differs a lot from M51. The latter shows discrete spiral arms while the former consists of small spiral arcs, hence its name: the Sunflower Galaxy. M63 is comparable in size and brightness to our galaxy. And there are even more similarities. Our galaxy belongs to a group, too, the Local Group. And the Milky Way has a large companion also, the Andromeda Galaxy, M31. It is possible that the Milky Way and M31 look similar to our view of M51 and M63 when observed from a star within the Sunflower Galaxy.

Messier 101

Our next object is one of the largest and brightest spiral galaxies of the Messier Catalog, Messier 101 in Ursae Major. And yet, it remains one of the most overlooked objects in the sky because of the galaxy's poor surface brightness. How can that be?

The answer can be found in the galaxy's large apparent size. M101 is seen face-on, and its galactic disk measures about 20' across. So the object's apparent brightness of mag 7.7 is spread out over an area as large as 2/3 the diameter of the Moon's disk. The result is that most of the M101's elusive light is lost in the light pollution of the sky. Chances are great that you might only be able to glimpse the galaxy's central bulge. Binocular observers have the advantage of observing at low power. Their instruments don't spread out the galaxy's disk too much in the field of view. That's why you should try to find M101 with your binoculars first. A telescope might spread out the feeble glow of M101 over too large an area for your eyes to detect it against the background glow.

Locating this giant pinwheel of stars is fairly simple. Let's move back to the Dipper's handle in Ursa Major, as shown in Fig. 6.1. On this finder chart, you'll see the location of M101. It's nearly halfway between Mizar and θ Boötis (see also Fig. 1.4, where Boötes and Ursae Major are shown in their full size). Our starting point is Mizar. From here, a chain of four bright stars

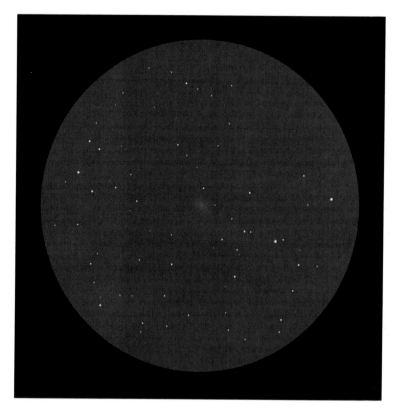

Fig. 7.4 Messier 101, the Pinwheel Galaxy, observed with observed with a pair of 8 × 56 binoculars

runs eastward over a distance of 4½° in the direction of θ Boötis. You can follow this trail of stars easily with your binoculars. Pay attention when you arrive at the fourth star. Messier 101 is only one and a half degrees away, to the NE. Look out for a soft, amorphous cloud measuring roughly 20′ across. Figure 7.4 shows what you might expect to see with your binoculars. The third and the fourth star of our chain of stars are located near the left side of the sketch.

Bear in mind that the contrast between M101 and its background depends on the darkness of your sky. In a light polluted area, the galaxy might not be visible at all. Don't give up immediately if you can't see the galaxy at first glance. At first sight, M101 is nothing more than a subtle brightening of the background sky. If necessarily, try to bracket your binoculars and give your eyes plenty of time to adapt to the view in the eyepieces. A pair of 8×56 binoculars can show a very diffuse but distinctly oval halo, measuring 22' across. The center of M101 appears slightly brighter. When you use a telescope, check what magnification works best for the halo and the central bulge of M101. On a very dark night, a small telescope can reveal several brighter features within the disk of M101.

Messier 101 is a very large spiral galaxy, at about 22 million ly away. With a diameter of roughly 190,000 ly, it's almost twice as large as our galaxy. M101 is remarkable for its large amount of stellar nurseries. Over 3,000 star-forming regions have been picked up by large telescopes. The galaxy's extreme activity is thought to be related to the tidal forces induced by a close pass by one of the companion galaxies of M101. Messier 101 is the brightest member of the so-called M101 group of galaxies. The galaxy's spectacular properties come to mind when we imagine the Milky Way placed next to M101 in space. Our galaxy would shine at mag 8.6 and measure less than half as large as the mighty Pinwheel Galaxy.

Arcturus

The final objects for this month's tour are located within the premises of the Milky Way. Let's first have a look at the brightest stars of spring evenings, Arcturus, which is in Boötes.

Locating Arcturus is straightforward if you know where the Big Dipper is. Take a look at Fig. 7.5, where the constellation of Boötes is displayed. If you follow the curvature of the handle of the Big Dipper towards the SE, you come to Arcturus. Imagining that Boötes resembles a kite also helps to distinguish the constellation among the other stars in the sky.

For us northern hemisphere observers, Arcturus is, after Sirius, the second brightest star in the sky. Arcturus has an apparent magnitude of −0.05. We inherited the name Arcturus from the Greeks, who called it the Bearkeeper. However Arcturus has played an important role in many ancient cultures.

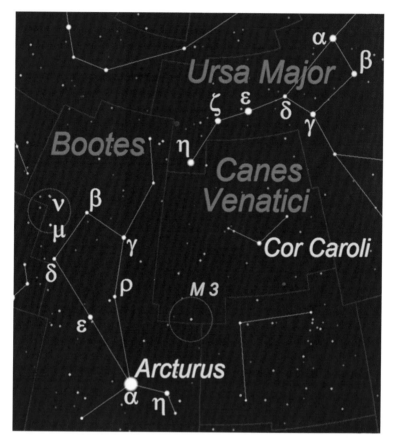

Fig. 7.5 Finder chart for Arcturus, μ and ν Boötis and Messier 3

The Polynesian navigators called it 'Hōkūléa,' the Star of Joy. They used the star Arcturus to navigate to the Hawaiian Islands. When a ship sails at the exact latitude of the Hawaiian Islands, the Star of Joy passes directly overhead.

Arcturus is an interesting star in many ways. It was one of the first stars of which proper motion was discovered. In 1718 the English astronomer Edmond Halley came to the conclusion that Arcturus was not a 'fixed' star. The discovery of the proper motion of stars was in fact a combined effort by two astronomers from totally different eras. Halley's conclusion was based on the precise position measurements of the stars by the Greek astronomer Hipparchos, roughly 1,850 years earlier. The proper motion of Arcturus is 2.29 arcseconds per year.

All stars are in orbit around the center of the galaxy. Stars closer to the center orbit at a higher speed than stars further away from the center. As a result, the positions of the stars in the Milky Way are not fixed. The constellations as we know them do change over time. The stars closest to our Sun are likely to display the largest proper motion. And indeed, Arcturus proves to be a close neighbor of the Sun. It is only 37 ly away (see also Fig. 4.2).

Arcturus' close distance is also one of the reasons for its high visual brightness. The other reason is that Arcturus is 100 times brighter in visible light than our Sun. It is 26 times the diameter of our Sun and has a surface temperature of 4,290 K. Arcturus is an orange K-type giant star in a late stage of its life cycle, and it contains about 1.5 times the mass of our Sun.

Astronomers are still debating Arcturus' origin. This orange giant differs a lot in composition from the mainstream of disk stars, which leads to the suggestion that Arcturus formed in a metal-poor era of our galaxy. Another unusual aspect is that Arcturus' orbit around the galaxy is highly inclined. Its lagging movement, compared to the disk stars, could be explained by two theories. One says that Arcturus might belong to an older population of our Milky Way. A more fascinating theory says that Arcturus belonged to another galaxy that merged with the Milky Way some 8 billion years ago. Proof could be found in the existence of the so-called Arcturus Stream, a collection of stars that share Arcturus' motion around the center of the galaxy.

In light of this theory, Arcturus' orange blaze in the late spring sky could represent the closest encounter humankind will ever witness with an exotic star of from an outlying galaxy. As a matter of fact, Arcturus has arrived at its closest point to us on its path through space. We see the star thus at its brightest. A million years ago, Arcturus' brightness was well below naked-eye visibility. It will take another million years for Arcturus to fade away again in the interstellar background glow. But for now, Arcturus shines bright and high during spring and summer evenings (Fig. 7.6).

Fig. 7.6 Arcturus, observed with a 8×40 pair of binoculars. Our Sun would shine at mag 5 when placed next to Arcturus. The fov is 8° wide

Boötes' Finest Double Stars

Now that we turned our gaze towards Boötes, why not hunt down some of its finest double stars. Our starting point for this star hop is δ Boötis, shown in Fig. 7.5. Delta itself is a fine but hard to split double star. The primary is a mag 3.5 G-type yellow giant at a distance of 117 ly. It is about 11 times larger than the Sun. The secondary is a mag 8 G-type main sequence star. It is just a little bit smaller than our Sun. It is possible to split Delta with 10×50 binoculars if you have a steady hand and sharp optics. Both stars are about 2' apart, but the secondary's glow might get lost in the vivid glare of the primary.

From δ Boötis, it's a 4½° jump to the NNE, where our next double star, fourth magnitude μ Boötis, begs for our attention. Mu Boötis is an F-type giant at a distance of 120 ly (see also Fig. 4.2). Take a careful look, as μ's companion lies only 1.8′ away. Binoculars with sharp optics should have no problems with splitting this fine double star. Telescopic observers could try to resolve the secondary, for it's in fact a close double star itself. Both companions are G-type stars, like our Sun. They are only 1.5″ apart.

The final pair for observation here is ν Boötis, a wide and thus easy pair for binoculars, which lies less than 4° NNE of μ. As shown in the sketch of Fig. 7.7, μ and ν Boötis fit properly in the same binocular field.

Fig. 7.7 Boötes, stars μ and ν, observed with a 8×40 binocular. The fov is 8° wide

Unfortunately, ν is not a real binary star. Its components are not physically related. They just happen to lie in the same line of sight. Both stars also offer fine color contrast. The westernmost component is a K-type orange giant, lying at 838 ly away, while the easternmost star is a white A-type dwarf star, 388 ly away. Keen-eyed observers could try to split ν with the unaided eye under a dark sky, as both components are of 5th magnitude.

CHAPTER 8

JUNE

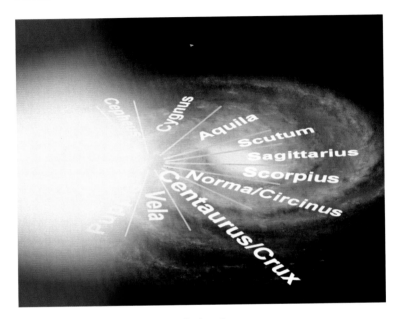

Fig. 8.1 Our night-time window during the summer season

The majestic constellations that we can see after dusk during this month herald the mighty rising of the summer Milky Way. Indeed, typical constellations for early June nights are Hercules, Lyra, Ophiuchus and Scorpius. The simple reason for this phenomenon is the slow orbit of our planet around the Sun. With it, our night-time window gradually shifts to the east. This month, the Sun passes in front of the constellations of Taurus and Gemini. The Sun also crosses the galactic plane of the winter Milky Way. As such, our night-time window points in the direction of the summer Milky Way. Figure 8.1 illustrates our night-time window during the summer season. The bright area reflects the constellations that are hidden by bright daylight. The summer season offers us an exciting vantage point to study

R. De Laet, *The Casual Sky Observer's Guide*, Astronomer's Pocket Field Guide, DOI 10.1007/978-1-4614-0595-5_8, © Springer Science+Business Media, LLC 2012

the feature rich innermost part of our galaxy, as it passes high overhead during the balmiest of nights.

This month we will encounter a new type of deep-sky object: the globular clusters. Globular clusters are the densest collections of stars within the Milky Way. They are spherical groups of hundreds of thousands of stars, firmly held together by gravity. Unlike the majority of stars that orbit in the galactic plane of the Milky Way, globulars can be found anywhere within the halo of our galaxy. Globular clusters orbit around the core of the Milky Way on eccentric elliptical paths similar to satellites. A globular cluster gains speed when it falls towards the core of our galaxy. It shoots at high speed through the galactic plane when it's the nearest to the center of the galaxy, and then slows down while it gains height again above the other side of the galactic plane. Compare it with a comet on its eccentric path around the Sun.

Another aspect that sets globulars apart from the majority of stars in the galactic plane is their composition and thus their origin. The stars that make up globulars are so-called 'metal-poor' stars, while stars like our Sun are 'metal-rich' stars. Astronomers believe that globulars are therefore the oldest building blocks of the Milky Way. They must be much older than the stars that make up the galactic plane of the Milky Way, because they formed in a metal-poor era of the universe.

A total of roughly 150 globular clusters have been discovered in the halo of the Milky Way. All galaxies seem to be populated with globulars. The Andromeda Galaxy, M31, harbors about 500 globulars. M87, a giant elliptical galaxy in Virgo, has a system of roughly 13,000 globular clusters in its halo.

The origin of globular clusters is still in debate. Some say that they are nothing more than the stripped-off remnants of cannibalized dwarf galaxies, pulled into the gravitational grasp of their more massive equivalents.

When you take a look at Fig. 8.2, you'll understand why June is an appropriate month to hunt down some of the most interesting globular clusters. Globulars are relatively small, and the nearest ones are still thousands of ly away from our Sun. This month, our early night-time window is pointed right above of the plane of the Milky Way, offering us a clear view on the

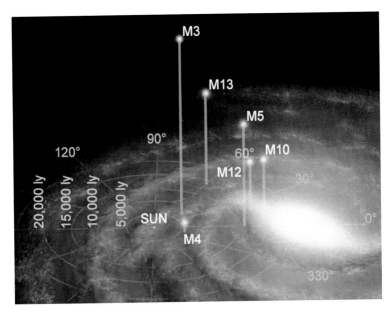

Fig. 8.2 The positions of the featured globular clusters within the northern halo of our galaxy, in relation to the galactic plane. Neither the sizes nor brightness of the globulars are drawn to scale

innermost part of the northern halo of our galaxy. It is here that we have the biggest chance of coming across globular clusters.

Fortunately, some of the brightest globulars are so conveniently placed in the June night sky that they are a true delight to be observed with even modest binoculars. Globular clusters resemble the look of faraway snowballs drifting through space. Comet hunters of old and present have often been fooled by these clusters that masquerade as tailless comets. For this very reason, the famous French comet hunter, Charles Messier, created his catalog of confusing deep-sky objects. The Messier Objects are still very popular among amateur astronomers today, hence their appearance in this guide.

The Pole Star and the Engagement Ring

Before we embark on a trip to the mysterious globulars that encircle the center of our galaxy, let's have a look in the opposite direction. Why not have a go at Polaris, the North Star? Now that twilight lasts so long, Polaris is an excellent object with which to start an observing session. It can be located even before night sets in, and it happens to be in very good company (Fig. 8.3).

When you train your binoculars on Polaris, you'll notice that it is part of a 40' wide ring-shaped asterism, called the Engagement Ring. It doesn't take

Fig. 8.3 Polaris and the Engagement Ring, observed with a pair of 8×56 binoculars

much imagination to see how Polaris represents the glittery diamond, held by this ring of sixth to ninth mag stars. The Engagement Ring is an excellent object to show to friends and relatives. They all will be delighted by this jewel in the sky, which can be seen on every clear night of the year.

When you have read chapters one and two, you will know how to find Polaris and you will understand how it is related to the galactic plane of our galaxy. But you should not forget that the North Star, too, is located within the disk of the Milky Way, as is shown in Fig. 4.2. Polaris is only 430 ly away. It's an evolved F-type supergiant with 45 times the diameter and 4 times the mass of our Sun.

Polaris has two companions. One is a ninth mag F-type main sequence star (dwarf) 18″ from the second mag primary. You can try to see Polaris' pale companion if your telescope can magnify 60–80 times. A third and very close F-type companion has been detected by the Hubble Space Telescope, making Polaris a triple star.

Our Sun, a G-type dwarf star, is no match in brightness to the Pole Star. Our Sun, when placed near the Pole Star, would shine with the brightness of a tenth magnitude star.

Messier 3

M3 is a fine example of a well-developed globular cluster. It is made up of roughly a half million stars, packed in a sphere as large as 190 ly. The combined luminosity of all its stars attains an apparent brightness of mag 6.2 at a respectable distance of 34,000 ly from us. The total mass of this globular is roughly 800,000 solar masses.

As shown in Fig. 8.2, M3 is further away from us than is the center of the Milky Way. This globular travels on an eccentric path around the core of our galaxy in about 300 million years. M3's distance to the center of the Milky Way varies between 15,000 and 50,000 ly. In each orbit, the globular crosses the galactic disk twice. On such an occasion, the tidal forces of the disk stars try to pull away the lightest stars of the globular cluster. This effect is called evaporation. Globular clusters must be very stable systems in order to survive the tidal shocks they undergo when crossing the galactic disk.

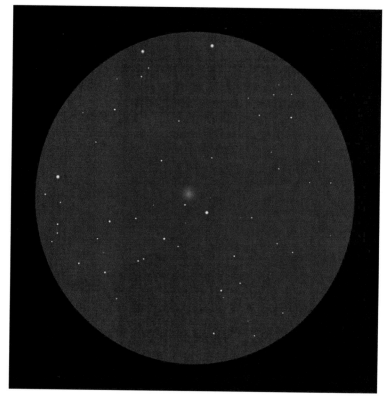

Fig. 8.4 Messier 3, observed with a pair of 8 × 56 binoculars

Because M3 is positioned very high above the galactic disk, we see it in the constellation of Canes Venatici. Figure 7.5 shows where M3 can be found. Locating Messier 3 with a pair of binoculars is very straightforward, as it lies somewhat less than halfway between Arcturus and Cor Caroli. What you see of M3 is very similar to what Charles Messier saw in 1764 with his imperfect instruments – a little nebula without any stars. M3's brightest stars are of mag 12.7, which is beyond the limiting magnitude of our small instruments. That is why we can only observe the combined glow of M3's unresolved stars as one amorphous cloudlet. It takes at least a 4-in. 'scope on a dark night to clearly resolve this globular's brightest Suns.

Figure 8.4 shows you what you can expect to see with your binoculars. M3's core appears bright and stellar. You can see it clearly with direct vision in an 8 × 56. The globular's halo is best appreciated with averted vision. While you admire the soft glow of this faraway city of light, imagine that our Sun would shine like a star of 20th magnitude at such a distance. Do you wonder how the night sky would look to an observer from a planet within M3's halo? Maybe the sky would be filled with so many Suns that there would be no such thing as a dark clear night at all. But if you were lucky enough to live in a solar system near the outskirts of M3, imagine how the Milky Way would unfold with its gracious spiral arms and brilliant core. But then again, we should not envy the extreme conditions within a globular cluster. Perhaps the hazardous radiation from all those densely packed stars makes life impossible. And think of the harmful passages a globular makes through the galactic disk, which could lead to a catastrophic ending for life on a planet in a globular. Maybe we should be thankful for the quiet and peaceful path our Sun follows in the galactic disk of our galaxy!

Messier 5

Another beautiful globular is Messier 5 in Serpens Caput, the Serpent's Head. This cluster lies closer to the core of our galaxy (see Fig. 8.2) than M3. M5 contains probably as many stars as M3, but in a sphere that is only 150 ly wide. This globular travels on a rather extended orbit, which can lead as far as 150,000 ly away from the core of the Milky Way. M5's orbital period is one billion years. Our distance to M5 is 26,600 ly. As a result, M5 appears a little brighter in our sky than M3 does. Messier 5 has an apparent brightness of mag 5.7. It would be a possible naked eye object if it were not placed so close to the fifth mag foreground star 5 Serpentis, which interferes with the sighting of the globular.

A pair of binoculars is all we need to identify this cluster as a non-stellar object. The star hop to Messier 5 starts from α Serpentis. Novice observers can start with the seasonal sky map (Fig. 1.5) as it represents the large constellation of Ophiuchus, the Serpent Holder. West of Ophiuchus and east of Boötes is the constellation Serpens Caput, the Head of the Snake.

Once you can locate Ophiuchus, use Fig. 8.5 to find α Serpentis. From here, slew your binoculars 8° to the SW until 5 Serpentis comes into view.

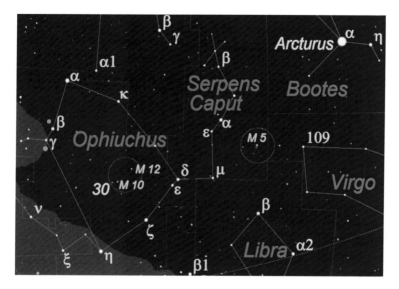

Fig. 8.5 Finder chart for Messier 5, Messier 10 and Messier 12

M5 lies less than half a degree NNW of 5 Ser. A pair of 8×56 binoculars clearly show the subtle glow of M5's extended halo, which gradually brightens towards its star-like core. A 3-in. telescope is capable of resolving individual stars as bright as mag 12.2 in this globular (Fig. 8.6).

M5's age is estimated at 8–10 billion years. This is, however, in conflict with the presence of a number of massive blue stars. These are the so-called 'blue stragglers.' More massive stars have a much shorter lifespan than the rest of the clusters' stars, and they should be the first to die out. The explanation for the existence of blue stragglers in such an old cluster could come from the close interaction between stars in globular clusters. One possibility could be that two stars merge into one more massive star with a higher temperature. Another theory says that a red giant star could have its expanding outer layers transferred to a close companion. This companion grows more massive to finally become a blue straggler. All globulars seem to have a number of blue stragglers, probably resulting from the large number of potential interactions in such densely packed clusters.

Fig. 8.6 The globular cluster Messier 5, observed with a pair of 8×56 binoculars

Antares and Messier 4

By the end of June, there is a particular star that is in opposition with the Sun. It rises when the Sun sets and vice versa. The star is the brightest star of the prominent constellation Scorpius, and it is called Antares. Scorpius is located near the southern summer Milky Way (see Fig. 1.5). Mid-northern observers need a clear flat horizon to the south to see any of the prominent stars that make up the Scorpion. From 51°N the tail of the scorpion never rises above the horizon. Figure 8.7 shows the bright sprinkling of stars

Fig. 8.7 Antares and the head of the Scorpion, sketched from an observation site at 51° north latitude. The tail of Scorpius is invisible from this latitude

that do rise high enough to be seen. The bright orange star is Antares. It represents the heart of the Scorpion. The arc of bright stars right of Antares represent the head of the Scorpion. The constellation is shown in full in the finder chart (see Fig. 8.8).

Scorpius is not a chance alignment of stars. Many of its stars, including Antares, are part of the Scorpio-Centaurus Association. This group consists of mainly young, hot blue main-sequence stars that all share the same motion through space. The Scorpio-Centaurus Association marks the inner edge of our local spiral arm, the Orion Spur. These stars thus lie in front of us when we look in the direction of the center of our galaxy (see Fig. 2.2).

Antares is the only star of the association that has already left the main sequence. It is an M-class red supergiant with a typical cool surface temperature of 3.600 K. Because the star lies so close to the ecliptic, it is

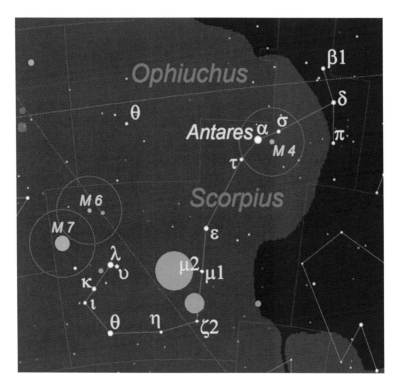

Fig. 8.8 Finder chart for M4, M6 and M7

often occulted by the Moon. Antares' peculiar orange color was already noted by our ancestors. Both the planet Mars and Antares have an orange hue, and the star was often mistaken for the planet. The Greeks called it the rival of Mars (Ares), hence the name Antares. We can assume that Antares once was a hot-blue O-type giant with a mass of more than 18 solar masses. When the star exhausted the hydrogen supply in its core, it turned into an expanded red supergiant.

Antares lies at 550 ly from us, which makes it the closest red supergiant to the Sun. It shines about 10,000 times brighter than the Sun in visible light. In other words, the Sun would appear as a star of 11th magnitude when placed next to Antares in the sky. If we take the infrared emission into account as well, Antares radiates roughly 60,000 times stronger than our day star. An object with a rather cool surface temperature can only exhibit such a strong energy output if it has an enormous facade. Consequently this red supergiant must have a diameter of about three times the distance from Earth to the Sun! If our Sun had this size, it would swallow all the inner planets, Earth and even Mars. The doom of Antares is to end its existence as a bright supernova somewhere between tonight and a million years to come. Earth is luckily far enough away to survive Antares' catastrophic ending. The blast will eject most of the star's debris into space, where it will be recycled into new stars and planets. Our Solar System, too, owes its existence to the heavy elements – enriched dust and gas – that once belonged to long-gone exploded stars.

The closest red supergiant shines with a fascinating orange glow, which is prominently obvious in any pair of binoculars. Antares also leads us to the closest globular cluster, cataloged as Messier 4. Figure 8.9 shows you how you can find M4. Just point your binoculars at Antares and you will also have M4 in your field of view. The soft glowing snowball one and a half degrees W of Antares appears to be a close neighbor of the red supergiant. But don't let this sight fool you. Antares, together with all the field stars, are mere foreground objects. M4, the closest globular, is at a distance of 5.640 ly, still much farther away. Figure 8.2 shows that M4 hovers roughly 2.000 ly above the galactic plane, in the direction of the central hub of the galaxy.

If you imagine a straight line between M4 and the Sun, remember that Antares and the Scorpio-Centaurus Association are only one tenth of that distance from the Sun. M4 shines with the brightness of a mag 5.8 star. This globular is a candidate for a naked-eye sighting, but the luster of the nearby Antares coupled with the low altitude of our target can make it a tricky task. Because of the cluster's proximity to the galactic plane, a considerable amount of the globular's light is absorbed by interstellar matter in our line of sight. Messier 4 is a rather loose globular without the typical densely packed core. A total of about 100,000 stars are estimated in a sphere that measures only 57 ly across. This globular is believed to be 10–12 billion years old.

M4 was the only globular cluster Charles Messier could resolve. Its brightest members form a particular star chain that crosses over the surface of

Fig. 8.9 Antares and Messier 4, observed with a pair of binoculars 15×70. The fov is 4.4° across

the cluster from north to south. These stars are of 11th magnitude. Unfortunately the low power of binoculars does not allow the sighting of these individual stars. A 3-in. telescope is able to show this facet. Nevertheless, from a dark site you might be able to locate the star chain with a pair of 15×70 binoculars. None of M4's stars are visible, but the brightened bar across the globular is prominently visible, as shown in Fig. 8.9. The globular's halo measures 15′ across. In a pair of 7×50 binoculars, this globular can be seen like a round soft glow with a diameter of roughly 10′.

Messier 10 and Messier 12

Not so far from M5, a nice pair of globular clusters awaits the binocular observer. Charles Messier included these globular clusters as the tenth and twelfth entry in his famous catalog. M10 and M12 are shown on the finder chart (see Fig. 8.5). You can again use α Serpentis to start with. Then follow the body of the snake through ε and η to δ Ophiuchi, near the western border of the Serpent Bearer. δ is escorted by ε Ophiuchi. This duo represents the hand of Ophiuchus that holds the snake. From Ophiuchus' hand, move two binocular fields to the East until fifth mag 30 Ophiuchus appears on the eastern edge of the field of view. With 30 Ophiuchus positioned as in the sketch (see Fig. 8.10), M10 and M12 shine fraternally next to each other.

Enjoy the unique wide-field view of this pair of globulars. Most telescopes can't show both objects well in the same field of view, for the separation between M10 and M12 is more than 3°. Globular clusters may all look the same with a pair of binoculars, but the more you observe, the more differences you will note between them. Now that they are visible in the same field of view, M10 and M12 form an excellent pair to study more deeply. Comparisons like these are excellent ways of improving your observing skills. With averted vision, you should find that M10 looks a tad brighter than M12. A careful look also confirms that M10 has the largest halo of the two. When using direct vision, you will notice that M12 shows a softer core compared to M10. The latter definitely has a sharper-looking nucleus.

In reality, these globulars are still about 4,500 ly separated from each other. They orbit the core of the Milky Way on separate paths, and they just happen to be in close conjunction. As a result, we see them as close neighbors in our sky. Figure 8.2 shows where M10 and M12 are, relative to our galaxy.

Messier 10 shines at mag 6.6 from 24,800 ly distant. Its mass is estimated at 200,000 solar masses. M10's true size is 140 ly. The brightest stars within this globular are of 13th magnitude. Resolution of the cluster starts with 3-in. telescopes.

Messier 12 lies closer to us, at 21,000 ly. Its apparent magnitude is 6.8. The size of the cluster is 85 ly, and its total mass is about 250,000 solar masses. M12 seems to have a rather small orbit. It never travels further than 20.000 ly from the galactic center. M12's orbital period is approximately

Fig. 8.10 The globulars M10 and M12 with 30 Ophiuchus (low left), sketched while viewing with a pair of 8×56 binoculars. The fov measures $5.9°$

130 million years. Its brightest stars reach mag 12. The globular appears granular in a 3-in. 'scope, while resolution is possible with 4-in. aperture.

Messier 13

We end our tour with a grand finale – the most famous globular cluster north of the celestial equator, also known as Messier 13. Our last object is often called the Great Globular Cluster of Hercules.

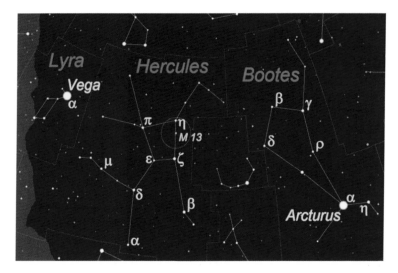

Fig. 8.11 Finder chart for Messier 13

M13 is very popular for several reasons. It's bright and high in the sky, easy to find and it's a great object no matter what instrument you use.

The constellation of Hercules represents the mythical Roman hero, the Strongman. Identifying Hercules is simple. This time of the year there is, besides Arcturus, one other very bright star in the sky. It's brilliant Vega, in the constellation Lyra, and it shines high overhead. The distance between Arcturus and Vega is 60°. Hercules fills in the space between Lyra and Boötes, as shown in Fig. 8.11. Try to locate the remarkable four-sided figure formed by π, η, ζ and ε Herculis. This quadrilateral is better known as the Keystone. Once you have found the Keystone, point your binoculars at η Herculis. Now gently move your view a mere 2° due south, towards ζ. Messier 13 should appear right in the middle of your sight. Figure 8.12 shows what you can expect to see. The bright star at the top of the sketch is η Herculis.

In a pair of 8×56 binoculars, Messier 13 looks like a conspicuous but fuzzy patch, flanked by a mag 7 star east and a mag 7.6 star southwest of the object. Notice how the brightness rises from the cluster's extremities to its

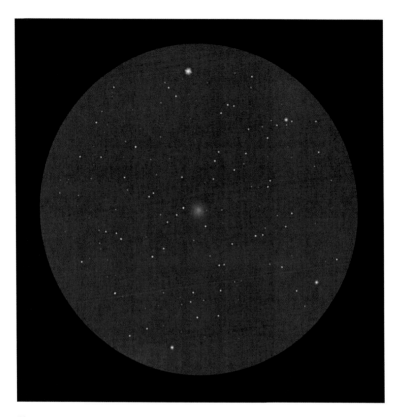

Fig. 8.12 Messier 13, observed with a pair of 8 × 56 binoculars. The fov measures 5.9° across

core. The latter may not appear star-like but rather granular. A mag 10 star is visible at the western edge of the cluster. With growing aperture, Messier 13 becomes a majestic sphere filled with thousands of tiny stars. A 4-in. refractor resolves many outliers that curve out of the granular core like long strings of minute stars. Between them, several dark wedges can be seen on the clearest of nights. When the atmosphere is very steady, the lowest power view at 18× makes these strings of stars hairy-looking. The view with a 10-in. telescope is simply breathtaking. Tens of thousands of stars are strewn together in an enormous ball of light.

Messier 13 is 26,000 ly away. It shines with the light of a mag 5.8 star. You may be able to identify M13 with the naked eye only at a remote site, when the sky is very dark. The nelm should be 6.0 or better for M13 to show up in the Keystone. This globular is 160 ly large and contains over one million stars. The brightest stars within M13 are of mag 11.9. The cluster's orbital period is about 500 million years. M13 can drift as far as 80,000 ly away from the galactic center. The globular's current position is illustrated in Fig. 8.2.

By now, you know how to find six great globular clusters. With a pair of binoculars, it is very easy to switch from one globular to another. Why not hone your observing skills by comparing these globulars with each other. See if you can notice differences in size, brightness and star concentration.

JULY

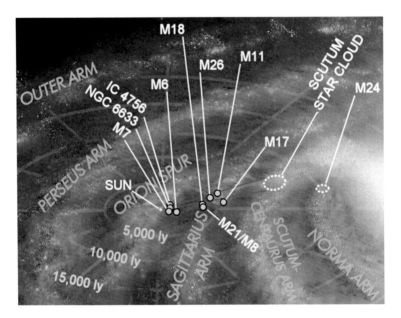

Fig. 9.1 The objects for this month all lie in the direction of the galactic center

Northern summer equals quality time for the deep-sky observer. Our night-time window is now pointed towards the target-rich environment of the summer Milky Way (see Fig. 8.1). Here we find the richest star clouds, clusters and nebulae of our home galaxy. The nights are the shortest of the year, but they are warmer, too. Summer is without question the most comfortable season for stargazers, both novice and experienced.

R. De Laet, *The Casual Sky Observer's Guide*, Astronomer's Pocket Field Guide, DOI 10.1007/978-1-4614-0595-5_9, © Springer Science+Business Media, LLC 2012

The Milky Way itself is a marvelous naked eye deep-sky object since it arches over our heads like the backbone of the night. From a dark observing site, the subtle glowing streamers of light from billions of stars is a breathtaking sight for the unaided eye. The Milky Way is a patchy and occasionally curling mass of light that's often interrupted by the darkness of nearby dust clouds, of which the Great Rift is the most obvious example. A detailed study of the structure of the summer Milky Way with the naked eye is described in Chap. 2. For now we will visit some of the most beautiful places that are within reach of binoculars and small telescopes (Fig. 9.1).

Messier 6

An attractive galactic cluster is positioned north of the tail of the Scorpion. It is known as Messier 6, the Butterfly Cluster. As shown in Fig. 8.8, M6 is easily found at 5° NNE of λ Scorpii, the star that marks the sting of Scorpius.

A dozen stars are visible with direct vision. Another 15 stars can be discerned using averted vision. This bright galactic cluster displays a startling form in binoculars – a butterfly with its wings wide open. Even the insect's antennae are present. The butterfly figure consists of mainly mag 7 to mag 8 stars. M6 at mag 4.2 is one of the brighter objects. The cluster is about 15' wide, which corresponds with a true size of 10 ly at a distance of 1,600 ly. M6's estimated age of 80 million years reflects the relative youth of this cluster. As many as 120 stars are located in the cluster. Most of them are hot blue B and A main sequence stars. The lucida of the cluster is, however, a K-type orange giant, called BM Scorpii. This variable star can influence the appearance of M6 with its brightness range between mag 5.8 and 8. BM Scorpii is in a later phase of its evolution. Its core has consumed most of its hydrogen. The star has left the main sequence and turned into a supergiant (Fig. 9.2).

Messier 6 is not a member of our spiral arm. Although close to the galactic plane, it is placed in the space between the Orion Spur and the Sagittarius Arm. Too bad that the cluster appears so low near the horizon. M6 would otherwise shine much brighter, and it would be a better appreciated naked-eye object.

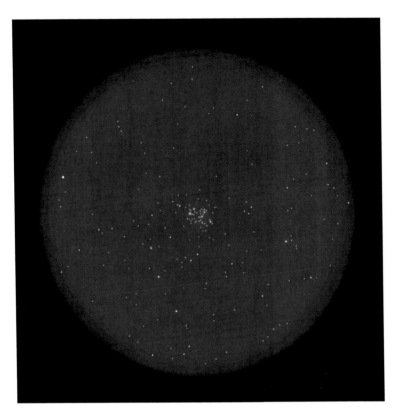

Fig. 9.2 Messier 6, the Butterfly Cluster, observed with a pair of 15×70 binoculars. The fov is 4.4° across

Messier 7

Only one binocular field SE of M6 lies the much larger cluster, Messier 7. The finder chart in Fig. 8.8 illustrates the positions of both clusters. With M7, we have reached the most southerly Messier object. From mid-northern locations, M7 only culminates ten degrees above the horizon. Observers north of 55° latitude never see this galactic cluster unless they travel south. The sketch of it in Fig. 9.3 was made at 39° latitude, where M7 and our previous target M6 show up as bright and crisp clusters.

Fig. 9.3 Messier 7, observed with a pair of 15×70 binoculars. The fov is 4.4° wide

Messier 7 at mag 3.3 is the third brightest Messier object, after M45 and M44. Because of its magnitude, it was known since antiquity as a fuzzy spot following the sting of the Scorpion. According to ancient Arabian scripture, M7 and M6 were called the Venom of the Scorpion. Sadly for us northern hemisphere observers, much of M7's luster is dimmed by the thicker atmosphere we have to look through when we gaze at objects near the horizon.

Messier 7 still belongs to the Orion Spur, as its distance is 1,000 ly. The cluster measures 23 ly across, which matches an angular size of 80′ in our sky.

A total of 750 stars are believed to belong to Messier 7. The cluster's lucida is a mag 5.8 G8 orange giant. The age of M7 ranges from 160 to 220 million years.

Messier 7 is so large and loose that it is best appreciated in low power instruments such as binoculars. Fifteen bright stars are scattered over an area of 80′. A bright double star can be seen near the center of the cluster. With averted vision, the number of cluster members can grow to over 50, depending on the size of the binoculars used. From our vantage point, M7 lies in front of a bright section of the central bulge of the Milky Way (see Fig. 9.1). A small number of the stars in the field around M7 don't belong to the cluster but rather to the inner spiral arms and the central bulge of our galaxy.

Messier 8, Messier 20 and Messier 21

Our next destinations lie in the constellation of Sagittarius, the Archer. Sagittarius is a real treasure chest for the deep-sky observer. Here we look deep into the inner structure of our own galaxy.

The constellation of the Archer lies east of Scorpius, as shown in Figs. 1.5, 2.10 and 9.4. A part of Sagittarius' stars are known as the asterism 'the Teapot.' λ Sagittarii forms the tip of the lid, while σ Sagittarii is part of the handle. The spout of the teapot is pointed towards M6. It is amusing to think of the Milky Way as the rising vapor coming out of the spout of the teapot.

Let's us first start with Messier 8, which lies 6°, or one binocular field, north of the tip of the spout of the Teapot. If you are lucky enough to observe from a dark location, then you can see M8 with the naked eye as a bright spot within the southern Milky Way (see Fig. 2.11). A pair of binoculars is all you need to discover the true nature of Messier 8, a stellar nursery that's brimming with newborn stars. The glowing nebula (mag 4.6) in which these new stars are born is called the Lagoon Nebula.

The view with a pair of binoculars is, to put it bluntly, spectacular. M8 appears as an elongated glowing cloud of gas, divided by a 'dark' river (the Lagoon), running from NE to SW. The eastern part of the nebula contains the very young open cluster of luminous blue O-stars called NGC

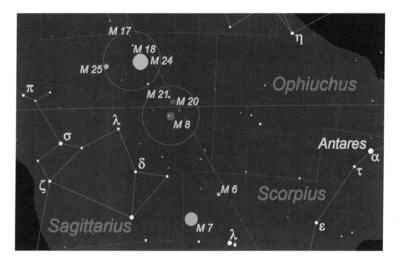

Fig. 9.4 Finder chart for the Sagittarius deep-sky objects

6530. This is, at mag 5.8, one of the brightest clusters within Sagittarius. About a dozen stars can be counted with 15 × 70 binoculars. NGC 6530 is 10 ly across.

The western part of the nebula harbors only two bright stars, together with a small bright patch of light, the Hourglass Nebula. This small patch looks like an out-of-focus star in a pair of binoculars. It is the brightest feature of the Lagoon. The brightest star in the western part of M8 is the mag 6 O4 star 9 Sagittarii. 9 Sgr is an enormous powerhouse that shines with the equivalent of 1.6 million Suns. The strong UV radiation of these powerful O stars ionizes and 'lights up' the molecular clouds of the Lagoon. Our Sun would shine with the light of a mag 15 dwarf when placed next to 9 Sgr. The Hourglass Nebula is the fuzzy star SE of 9 Sgr. Larger telescopes will reveal its true shape.

The Lagoon has more treasures to offer if you continue looking. The patient observer will be rewarded with the view of fainter nebulous extensions and delicate curls of dark lanes. Altogether, the Lagoon Nebula is a very complex star-forming region at a distant of 4,300 ly in the Sagittarius spiral arm of our galaxy. Its real size is approximately 110 × 50 ly. M8 is part of a larger formation, called the Sgr OB1 Association. The Lagoon

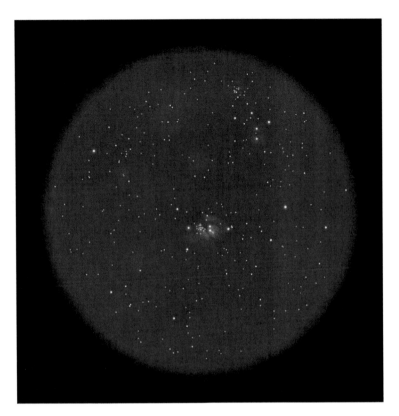

Fig. 9.5 M8, M20 and M21, observed with a pair of 15×70 binoculars. The fov measures 4.4° across

Nebula is perhaps the most beautiful nebula of the summer Milky Way. Its counterpart in the winter Milky Way is the Orion Nebula, which lies in the opposite direction of the Scorpion.

Just 400 ly closer to us and 2° north of M8 lies the open cluster Messier 21. It also belongs to the Sgr OB1 Association. M21 is 20 ly in diameter and consists of roughly 100 stars. Its brightest members are young B stars, giving the cluster an age of 4–8 million years. Small binoculars show a minute concentration of half a dozen stars in hazy surroundings. M21's brightness is of mag 5.9. The cluster is located at the upper edge of the sketch in Fig. 9.5.

When the night is very dark, you can try your luck at observing Messier 20, the Trifid Nebula. M20 lies 35' south and a little to the west of M21. Here you see a collection of five brighter stars. A pair of 15×70 binoculars reveals a soft halo around the two southernmost stars. This duo of stars is embedded in an extremely faint emission nebula. M20, which lies at a distance of 2,600 ly, is also a star birth region similar to M8. The most southerly star is in fact a multiple star system with an youthful age of not more than 400,000 years. The dark nebula, Barnard 85 (B85), that divides M20 into three lobes, is not that difficult to see. You may be able to see M20 together with B85 using a 4-in. telescope equipped with a special nebula filter (UHC) in front of the eyepiece.

The five stars together with M20 and M21 is sometimes referred to as the asterism Webb's Cross, with M21 is at the top and M20 at the bottom of the cross.

IC 4756 and NGC 6633

Our next destinations lie on the border of Ophiuchus and Serpens Cauda. Here we find two beautiful galactic clusters, IC 4756 and NGC 6633, which form a delightful duo in a pair of binoculars. IC 4756 can be found at one binocular field west of θ Serpentis, the fourth mag star at the tip of the serpent's tail. When you place this cluster in the eastern half of your fov, NGC 6633 comes into view as well (Fig. 9.6).

Both clusters are reasonably bright objects of mag 4.6. They can be seen with the naked eye on a dark and moonless night. IC 4756 is the hardest to discern, because the cluster spans almost one degree in the sky. It is an evolved cluster with an age of almost 800 million years. The cluster lies 1,300 ly just beyond the edge of our Orion Spur. Its true size is about 20 ly. A pair of 8×56 binoculars show many ninth and tenth magnitudes pinpricks framed by a few brighter stars. The cluster comes into its own with averted vision. Then IC 4756 looks like a faint cloud of diamond dust.

NGC 6633 lies only 3° WNW of IC 4756. This cluster, at a distance of 1,000 ly, is placed at the edge of our local spiral arm. Its apparent size of 30' reflects a true size of 9 ly. NGC 6633 has an age of 600 million years. It is the

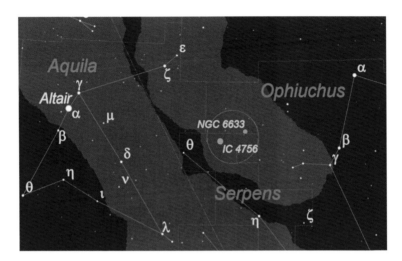

Fig. 9.6 Finder chart for IC 4756 and NGC 6633

smallest but also the most conspicuous of the duo. Four brighter stars are embedded in an elongated haze of fainter cluster members (Fig. 9.7).

We can see these clusters with binoculars because they have some very bright Suns among them. Average dwarf stars such as our Sun remain invisible in binoculars at such distances.

Messier 24

When Charles Messier added M24, the Small Sagittarius Star Cloud to his catalog, he believed that it was a collection of stars embedded in an enormous nebula. But M24 is not a deep-sky object in the true sense of the definition. Messier 24 is a unique window to the interior of our galaxy. Our view of the inner structure of the Milky Way is in great measure obstructed by the dense dust clouds within the spiral arms that curve into the central bulge of the galaxy. Even the small section of the Orion Spur that we look through is crowded with obscuring dust clouds. The Great Rift is the most obvious example of these. When we look at M24, we happen to see through

Fig. 9.7 IC 4756 and NGC 6633, observed with a pair of 8 × 56 binoculars

an unobstructed tunnel from within the Orion Spur, through the Sagittarius Arm and the Scutum-Centaurus Arm and into the incurving part of the Norma Arm.

The view spans a total distance of approximately 16,000 ly. The combined brightness of all the stars in this viewing tunnel is of mag 2.5. M24 covers an area of 120′ × 45′ in the sky. That's why the Small Sagittarius Star Cloud is clearly visible on a moonless night as a small cloud within the Milky Way. How detailed Messier 24 will look from your observing site depends largely on the darkness of your sky. The majority of the stars within M24 are too faint to be resolved with normal binoculars. As a result, M24 displays a soft background glow with nearby stars strewn over the whole area. There is a prominent asterism of a dozen mag 6 to mag 8 stars in an oval shape. The asterism reminds some of Captain Nemo's Nautilus from the Jules Verne's novel, *20,000 Leagues under the Sea*. The Nautilus comes straight on through the tunnel of M24. The Nautilus asterism is even visible in binoculars under light polluted skies.

If you are lucky enough to observe under a dark sky, look out for two dark spots just north of the Nautilus. These spots are two dark nebula, called Barnard B92 and B93. They probably are foreground objects within our Orion Spur. If B92 and B93 show up clearly, try to detect the two dark lanes emerging from the westernmost dark nebula, B93. On the clearest of nights, tracing out the dark frame of dust clouds around M24 is a pleasant exercise. Even a small pair of 8×40 binoculars is capable of showing all these features. The view with a pair of 70 or 100 mm binoculars is even more astonishing. The whole tunnel of M24 is filled with thousands of faint glittering stars that emerge out of the background glow. There is probably no other section visible from mid northern latitude where such a gathering of stars can be seen in a single field of view. Barnard 92 and 93 appear more prominently as well. They show up as dark eyes on top of the Nautilus, which suddenly turns into a large crab seen face-on with its black eyes looking at you. The Small Sagittarius Star Cloud and its surrounding may be the most impressive binocular showpiece for northern latitude observers.

A number of interesting deep sky objects are present in the immediate vicinity of the Small Sagittarius Star Cloud. All of them are foreground objects when compared to the window offered by M24. The open cluster M25 can be found at 5° east of M24. Small binoculars reveal a knot of 9 stars surrounded by haze. M25 lies 2,000 ly away and shines as bright as mag 4.6. Its diameter of 30′ corresponds with a true size of 17 ly. It is shown near the left border of Fig. 9.8. Another open cluster, M18, lies 1°50′ north of M24. In binoculars, a 5′ wide fuzzy area occupied by a half dozen stars is visible. Messier 18 lies at a distance of 4,200 ly in the Sagittarius Arm. The star cluster includes a number of blue supergiants, hence its estimated age of 50 million years. M18's integrated brightness is mag 6.9.

One degree north of M18 lies the beautiful Swan Nebula, Messier 17. This emission nebula with its brightness of mag 6 and its size of $30′ \times 20′$ is

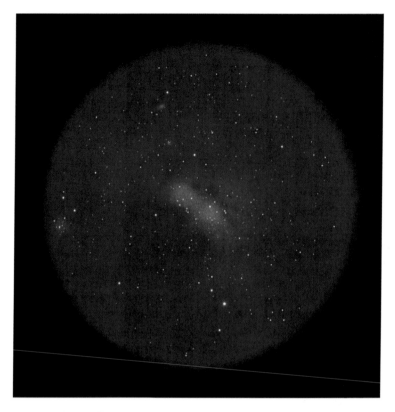

Fig. 9.8 The Small Sagittarius Star Cloud, Messier 24 observed with a pair of 8 × 40 binoculars. The fov is 8° wide

visible with a pair of binoculars, which show a subtle bar of light, like a comet tail. M17 is, just like M8, a stellar nursery. It too is located in the Sagittarius Arm. The Swan Nebula is 5,900 ly away and has a true expanse of 70 × 50 ly. Hidden in its cocoon lies a million year old cluster of over 2,000 fresh born stars. M17 is a bright and beautiful nebula that is definitely worth a look with the higher magnification of a telescope.

For the sketch in Fig. 9.8, a small pair of 8 × 40 binoculars were used. They offered a generous fov of 8°. With these little binoculars, you can have in your sights a window into the interior of our galaxy (M24), two dark nebula

(B92 and B93), two open clusters (M25 and M18) and one emission nebula (M17). No other instrument, no matter how powerful, could have presented this wonderful stretch of our Milky Way in a single view.

The Scutum Star Cloud

Because it rises higher in the sky than the Great and Small Sagittarius Star Clouds, one of the brightest star clouds for mid-northern observers is the Scutum Star Cloud. How the Scutum Star Cloud can be found with the naked eye is described earlier in this chapter. For now we will cover its binocular appearance. The Scutum Star Cloud lies about one binocular field WSW of the third mag λ Aquilae. It can be seen with the naked eye on a moonless night. This bright region of the summer Milky Way is in fact an unobstructed window to the incurving arc of the Scutum Centaurus Arm, about 13,000 ly away. Through this window we don't just look at but along an inner spiral arm. Therefore its cross section appears dense and bright. Because in this constellation the Great Rift is very prominent and north of the galactic plane, we only see the fragment of the incurving Scutum Centaurus Arm that lies south of the galactic plane (Fig. 9.9).

The look of the Scutum Star Cloud is shown in the sketch (see Fig. 9.10). Through a pair of 8 × 40 binoculars, the star cloud measures 5° × 3°. It is less densely packed with tiny stars than M24. In the NE, the Scutum Star Cloud is bordered by the dark nebula designated as Barnard 111. A bent chain of bright stars that run from λ Aquilae towards β Scuti (the bright star at the top of the sketch), crosses B111. Along this chain lies the famous open cluster Messier 11. Another star chain runs parallel from β Scuti to α Scuti (the bright star at the right edge of the sketch). A subtle dark lane that lies immediately SE of this star chain, seems to cut the Star Cloud in two. Another but more distinct dark lane that starts from the λ Aquilae region and runs towards δ Scuti has clearly separated the southeasternmost wedge of the Scutum Star Cloud. The way in which the Scutum Star Cloud is framed by nearby dust clouds, give it the shape of a medieval shield, hence the name of the constellation.

A number of binocular deep-sky objects are present in this interesting stretch of our Milky Way as well. The brightest object is Messier 11, a

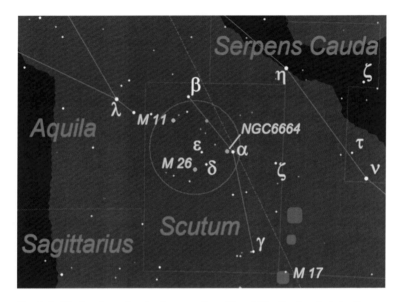

Fig. 9.9 Finder chart for the Scutum Star Cloud. The finder *circle* indicates its position

crowded but compact open cluster near the NE border of the Scutum Star Cloud. Binoculars show M11 like a single star in a fuzzy spot of 10′ wide that shines with the integrated brightness of a mag 5.8 star. Messier 11 is a very rich open cluster. A total of 2,900 stars are believed to belong to this galactic cluster, which has star count and density that rivals the specs of a loose globular cluster. It also is much better appreciated with the medium magnification of a telescope that will completely resolve this beautiful cluster. The size of M11 is about 30 ly. At a distance of 6,100 ly, the cluster clearly is a foreground object that resides in the Sagittarius Arm, just like M17. The age of Messier 11 is believed to be 250 million years.

The next brightest galactic cluster in Scutum is NGC 6664, at mag 7.8. This 15′ wide cluster lies no more than 30′ east of α Scuti. Observing NGC 6664 with normal binoculars is really demanding, as the bright glare of α Scuti interferes with the weak glow of the cluster. In my 8 × 40 binoculars, NGC 6664 could easily be mistaken for a ghostly reflection of α Scuti's radiance. This cluster measures roughly 30 ly across, and it resides at a distance of 6,200 ly in the Sagittarius Spiral Arm. A third galactic cluster in

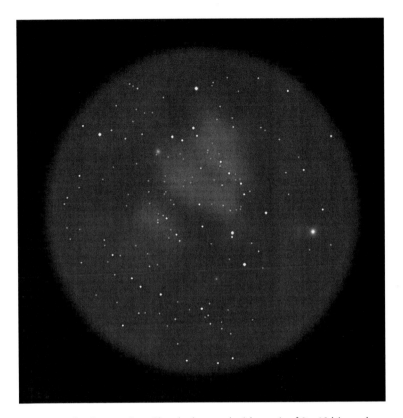

Fig. 9.10 The Scutum Star Cloud, observed with a pair of 8 × 40 binoculars. The fov measures 8° across

the foreground of the Scutum Star Cloud is Messier 26. Although this mag 8 open cluster measures only 8′ across, it is much easier to see than NGC 6664. M26 lies in a star poor area, 1° ESE of δ Scuti. A pair of 8 × 40 binoculars clearly show M26 appearing as an out-of-focus star. The cluster's size is 12 ly at a distance of 5,100 ly. Messier 26 is 90 million years old and a true tracer of the Sagittarius Spiral Arm.

The binocular observation of the Scutum Star Cloud reveals a true sense of depth in the eyepieces when we imagine that these three tracers of the Sagittarius Spiral Arm lie half as far as the incurving arc of the Scutum Centaurus Arm.

AUGUST

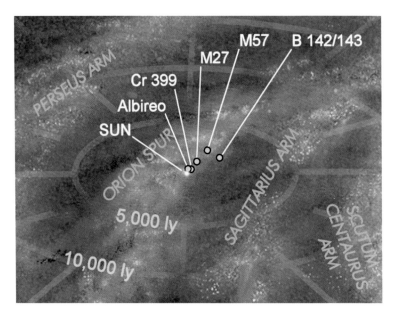

Fig. 10.1 The objects of this month all lie in the Orion Spur

In August, the evenings get darker sooner while they remain mild. Right after dusk, the magnificent Milky Way displays its complete splendor from Sagittarius in the lower southwest over Cygnus at the zenith to Cassiopeia near the northeastern horizon. The objects for this month all lie in the galactic front yard of our Solar System (Fig. 10.1).

A Tour Around Vega

The brightest star of the summer constellations is the luminary of the compact constellation Lyra (the Harp), better known as Vega. Vega is also the brightest star of the Summer Triangle (see Fig. 2.10). The name Vega is

R. De Laet, *The Casual Sky Observer's Guide*, Astronomer's Pocket Field Guide, DOI 10.1007/978-1-4614-0595-5_10, © Springer Science+Business Media, LLC 2012

derived from an old Arabic name for 'swooping vulture.' Vega is, at mag zero, the fifth brightest star in the sky, right after Arcturus. It's an A0-type main sequence star at a relative close distance of 25.3 ly (see Fig 4.2). Both Vega and the Sun are main sequence stars. A main sequence star fuses hydrogen into helium in its hot dense core. But Vega has 2.5 solar masses and is about 37 times more luminous than our Sun. At Vega's distance, our Sun would shine at mag 4.25.

A-type stars consume their fuel at a much higher rate than G-type stars, like our Sun (see Table 1.3). As a consequence, Vega's total main sequence lifetime, before it will become a red M-type giant, is one-tenth of our day star's. Vega is only 450 million years old and already halfway through its main sequence lifetime. It appears that Vega has an extremely short rotation period of 12.5 h (our Sun had a rotation period of 25 days). As a result, the star looks like a flattened sphere with a 2.26 solar polar diameter and an equatorial diameter of 2.78. If Vega would rotate any faster, it would literally fly apart. We like to associate Vega with the northern summer season, but this has not always been the case. As explained earlier in Chapter 2, the north celestial pole describes a circle in the sky. As such, Vega was our Pole Star around 12,000 B.C. And it will return to that status near A.D. 14,000.

Of all the naked-eye stars that lie near the precession circle, Vega is the brightest. Vega also roughly indicates the direction in which our Solar System moves through space in its orbit around the galactic center, while Sirius marks the spot where our Solar System originated. In a way of speaking, Sirius lies in our backyard and Vega in our front yard.

A-type stars such as Vega are hotter than the Sun and they shine with a bluish-white tint. See if you can distinguish with the naked eye Vega's icy blue hue from Arcturus' golden tinge. In a pair of binoculars, Vega's sapphire tint is evident. Additionally, in the wide field of your binocular eyepieces, a few other celestial gems accompany Vega. One of the most famous double stars in the sky, the Double Double ε Lyrae, just lies 1.5° NE of Vega. Your binoculars will show a binary star system with a separation of 3.5 arcminutes. ε1 is the northern component; ε2 is the name for the southern one. If you have keen eyesight, you can try to split ε Lyrae on a steady night without any optical aid. A small telescope at a power of 75× will split both stars again because the components of ε1 and ε2 are, respectively,

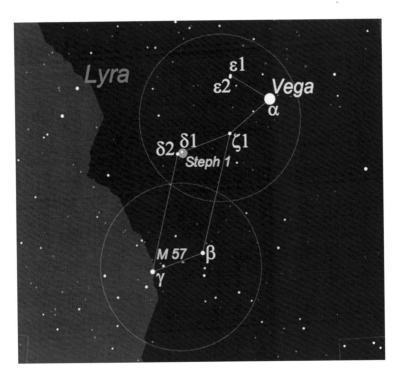

Fig. 10.2 Finder chart of the objects in Lyra

separated by 2.6″ and 2.3″. The four components all are A-type stars with more or less the same brightness (Fig. 10.2).

The Double Double lies at a distance of 160 ly. The binary ε1 has a rotation period of about 1,200 years, while the ε2 binary probably has half that period. ε1 and ε2 are separated by more than 0.16 ly. It is very doubtful whether they can remain gravitationally bound while traveling through the influence of passing stars within the galactic disk.

In the same binocular field, as shown on the sketch in Fig. 10.3, are the northernmost cornerstones of Lyra's parallelogram, δ and ζ Lyrae. δ Lyrae is easily split into the 10′ wide double, δ1 and δ2 (see also Fig. 10.2). However, they are not physically bound but a chance alignment in our line of sight.

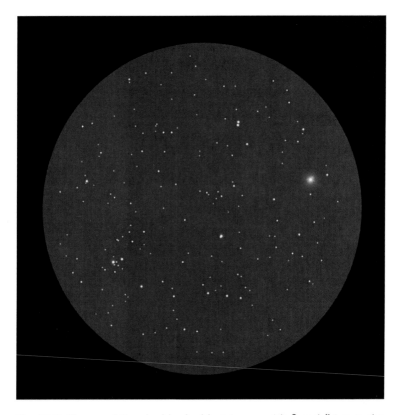

Fig. 10.3 Vega and the double-double ε Lyrae, with δ and ζ Lyrae, the northernmost corners of the parallelogram of Lyra. Observed using a pair of 8 × 56 binoculars. The field of view is 5.9° wide

The closest to us is δ2, with a distance of 900 ly. δ2 was a former hot B-class star with a mass of six Suns. It is now (at 75 million years old) a cool red giant of mag 4.5, with a radius of 200 times that the Sun's. And at 1,100 ly away at mag 5.5 lies δ1, a 35-million-year-young hot blue main sequence B-star, located at the heart of a small cluster called Stephenson 1. This young cluster measures only 6 ly across and counts about 15 members. In binoculars, Stephenson 1 can be seen as a loose gathering of mag 8 to mag 10 stars around δ1 Lyrae. Our day star would look like a mag 12.4 star when placed in the Delta Lyrae cluster.

Messier 57

Just one binocular field south of δ and ζ Lyrae appear the southernmost cornerstones of the Harp, formed by the stars γ and β Lyrae. These two stars also are the bright beacons that lead the way towards Messier 57, the famous Ring Nebula. M57 lies about halfway between γ and β Lyrae, as shown in Fig. 10.4. You might think that the sketch is 'empty,' but take a second look. There is a tiny puff of smoke drawn right in the center of the sketch. This is M57. When you aim your binoculars at γ and β Lyrae, look out for a fuzzy mag 8.8 star that shows up with averted vision but disappears

Fig. 10.4 Binocular view of M57 in between γ and β Lyrae. M57 is right in the middle of the sketch. The fov is 5.9° across

with direct vision. This effect is known as the blinking of the planetary. M57 is only 1 arcminute large and thus better suited for larger instruments.

The Ring Nebula might not look spectacular in our binoculars, but it is the archetype of a specific sort of deep-sky object, the planetary nebula. Planetary nebulae are the ejected outer shells of a star in its death throes. These nebulae have nothing in common with planets whatsoever. When these objects were discovered in the eighteenth century, they were mistakenly categorized as giant planets. The misleading term, however, is still used in modern astronomy.

The majority of the stars, including our Sun, will produce a planetary nebula at the end of their life. Planetary nebulae form a crucial phase in stellar evolution, because they are an important mechanism for mass transfer from stars to the interstellar medium. As such, planetaries feed the interstellar medium with enriched elements of which new stars will be created. Earth and our species with it owe their existence to the heavier atoms that once were forged in dying stars of our galaxy.

It would seem logical that we would find many planetaries within the galactic disk, since it is the fate of most stars. However, the contrary is true. Planetary nebulae are difficult to see because they are rather small and are dim emission nebula. They only glow because of the UV radiation they absorb from their central star. Planetaries exist for only a small amount of time because they keep expanding until they are completely diffused within the surrounding interstellar medium.

The Ring Nebula in Lyra is 2,300 ly away. It is lit up by the UV radiation from its central white dwarf that has a surface temperature of roughly 120,000 K. The Ring Nebula has a true diameter of 0.9 ly. It expands 1″ every 100 years. A small telescope will reveal the ring shape of M57. It takes a really large telescope to show the mag 15.6 central white dwarf of M57. The Ring Nebula was formed about 10,000–20,000 years ago. It will probably take as long before the nebula will be dissolved totally in space. White dwarfs are doomed to cool down until they vanish in the darkness of space. However, they are so hot, heavy and compact, that it takes longer than the age of the universe for them to radiate away all their energy. Therefore, the first stars in the universe that ever turned into white dwarfs are still hot, luminous bodies.

Messier 27

The most spectacular planetary nebula for amateur astronomers, Messier 27 can be found in the small constellation of Vulpecula. M27 is an easy binocular target because it lies only 3° north of γ Sagittae, as shown on the finder chart in Fig. 10.5 (see Fig. 1.5 to locate Sagitta). M27, the Dumbbell Nebula, is 1,150 ly away and has the integrated brightness of a mag 7.4 star. Its 5′ wide surface is therefore a more prominent disk in our binoculars than M57. With a pair of 15×70 binoculars, M27 starts to show a more pronounced rectangular shape (see Fig 10.6). You can even tell the orientation of the major axis of this nebula. A small telescope is all you need to see the beautiful dumbbell shape of this 5 ly-sized nebula. Nevertheless, we only see the brightest part of M27. The Dumbbell Nebula is veiled in a gossamer halo that's twice the size of the hourglass shape. The total size of M27 would stretch all the way from our Sun to Sirius.

The nebula expands at a rate of 2.3″–6.8″ per 100 years. M27's central star is a white dwarf of mag 13.5. It is the strong UV radiation of this cooling

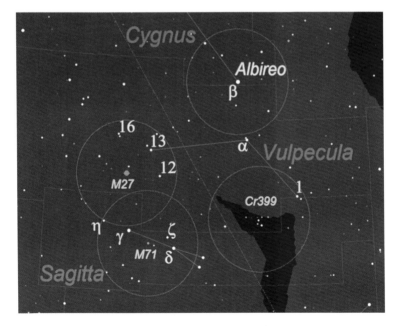

Fig. 10.5 Finder chart for M27, Albireo, Cr399 and M71

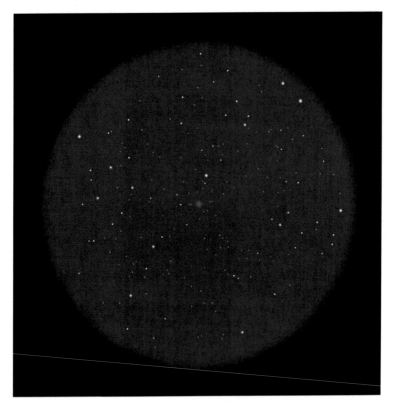

Fig. 10.6 Messier 27, observed with a pair of 15×70 binoculars. The fov is 4.4° across

white dwarf that causes the nebula to glow. M27 is about 9,000 years old and might take as long again to completely dissipate into space.

Albireo

Cygnus, the Swan, is a striking constellation that flies high overhead on August evenings. It is sometimes also called the Northern Cross. Some feel it more closely resembles the bird. Deneb marks the Swan's tail and Albireo represents the head. Thus the Swan flies in the direction of Vulpecula.

Albireo, or β Cygni, is one of the finest double stars for amateur astronomers. It is represented in the finder chart of Fig. 10.5 (see also Fig. 2.10). With a separation of 34″, Albireo's components are a real test object for higher power binoculars. With a pair of 8 × 56 binoculars, the pair can be split, but with difficulty. A pair of 10 × 50 bino's do a better job, while the 15 × 70 have no trouble at all with resolving Albireo into a golden yellow mag 3.1 primary (Albireo A) and a bluish mag 5.5 secondary (Albireo B). The perception of color is a personal affair.

A small telescope at 60× or more shows the beautiful color contrast of Albireo's components much better. Here is a little trick you can use. To better see star colors, slightly defocus your optics so that the stars appear as small disks. Bright objects, such as double stars and vivid open clusters, don't need dark skies or moonless nights to be observed. Take advantage of a clear, moonlit sky to sketch the bright stars of Albireo. The dark blue sky of Fig. 10.7 reflects the effect of the nearby full Moon.

The Albireo system, which lies at 380 ly away, is in fact a triple system. The golden primary, Albireo A, consists of a K-type orange giant with five times the Sun's mass and a very close hot blue B-type companion with three times the solar mass. They are too close together to be separated with a telescope. The blue secondary, Albireo B, is also a B-type dwarf with three solar masses. The stars of the Albireo system are all much brighter powerhouses than our Sun, which would shine at mag 10 when placed near them.

Collinder 399

A real treat for binocular observers is the asterism Collinder 399 in the constellation of Vulpecula (the Fox). With the use of the finder chart of Fig. 10.5, you can star hop 8° from Albireo in Cygnus over fourth mag α Vulpeculae towards Collinder 399 (Cr 399). A striking alignment of nine bright (sixth to seventh mag) stars, resembling the bar and the hook of a coat hanger, will show up in your eyepieces. (That's why C399 is known as the Coat Hanger.) Telescopic observers will have a hard time squeezing the 1.5° wide Cr 399 in their field of view. Cr 399 is at mag 3.6, a fine

Fig. 10.7 Albireo, observed with a small telescope at 62×, when the Moon was full and the sky dark blue. The fov is 50′ large

naked-eye object on a moonless night. Without optical aid, you can see Cr 399 as a hazy double star.

Collinder 399 was once cataloged as a galactic cluster at 420 ly away. However, recent observations have determined that Cr 399 is just an amusing chance alignment of stars. The Coat Hanger is one of the summer highlights to show with your binoculars to relatives and friends (Fig. 10.8).

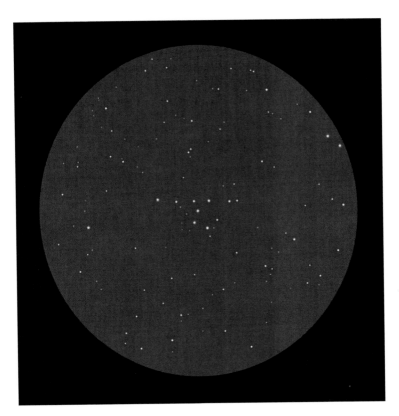

Fig. 10.8 Collinder 399, observed with a pair of 8 × 56 binoculars

Barnard's 'E'

One of the most frequently overlooked types of deep-sky objects are dark nebulae. These clouds are of the same nature as the reflection and emission nebulae we have come across before. They remain dark due to the lack of stars to light them up. Opaque interstellar clouds give away their silhouette when they lie in front of a bright stretch of the Milky Way.

That dark nebulae are beyond the reach of binoculars is a misconception. Even the naked eye is capable of tracing out the contours of the Great Rift.

Fig. 10.9 Finder chart for Barnard's E (B142-3)

Binoculars take in a lot of field and are most suitable to observe large dark nebulae. The only qualifiers are that your observation site is reasonably dark and that the nebula rises high in a transparent and moonless sky. If the Milky Way appears as clearly as in the sketch (see Fig. 2.9), then you'll have a good chance of spotting dark nebulae with your binoculars. One of the most obvious dark nebula is Barnard 142–143, next to Altair in the constellation of Aquila. Use γ and χ Aquilae as your signpost for this tour. Both χ and B142-3 lie 1.5° away from γ (Fig. 10.9).

Try to aim your binoculars like shown in the sketch (see Fig. 10.10) and give your eyes plenty of time to adjust. Keep bright Altair out of the field, as its bright glare will definitely ruin your valuable night vision. Use your averted vision to its best advantage and keep looking around in the eyepieces. The central EW elongation of the 'E' is the most prominent feature of the nebula.

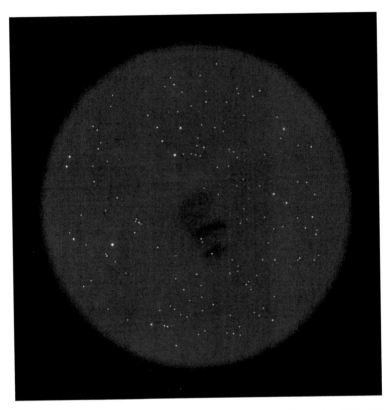

Fig. 10.10 The dark nebula, Barnard's E, observed with a pair of 15×70 binoculars. The fov measures 4.4° across

It might take a while before you can confirm the delicate presence of the E-shaped cloud in front of the bright Milky Way background.

Estimating the distance of such a cloud is a difficult task, but Barnard's E probably lies at 2,000 ly from us. This obscure cloud belongs to our local spiral arm, just like the Great Rift. It is in dense clouds like these that new generations of stars are born.

CHAPTER 11

SEPTEMBER

Fig. 11.1 Our night-time window to the Milky Way in autumn

The summer season ends this month. We not only notice this from changes in nature but also from the new constellations that rise in the east. The Sun now moves towards the autumnal point in Virgo, which will be reached around September 22 or 23. Our night-time window is slowly shifting away from the galactic center towards higher galactic longitudes (see also Fig. 2.3, where the star on the ecliptic indicates the position of the Sun around the autumnal equinox). The star-filled stretches of the Milky Way still dominates the sky. The deep-sky objects on the incurving side of our own Orion Spur come within reach of our binoculars and small telescopes on September evenings (Fig. 11.1).

R. De Laet, *The Casual Sky Observer's Guide*, Astronomer's Pocket Field Guide,
DOI 10.1007/978-1-4614-0595-5_11, © Springer Science+Business Media, LLC 2012

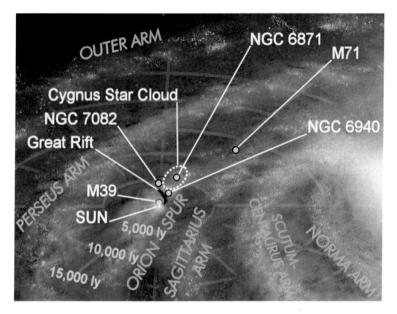

Fig. 11.2 The deep-sky objects of this month lie on the incurving arc of the Orion Spur

Messier 71

We already visited in the month of June a selection of bright globular clusters that all are located in the halo above the galactic plane. However, a few globulars are present in the star dense galactic plane as well. They are of course harder to observe because the opaque interstellar clouds within the spiral arms seriously hamper our view of the deep-sky objects within the galactic plane. Messier 71 is a globular cluster that follows an awkward path through the galactic plane. The globular's eccentric orbit never leaves the galactic disk of our Milky Way. M71 must therefore travel through several spiral arms for about 160,000 years to complete a full orbit. The poor cluster faces being disrupted due to the tidal forces exhibited by the stars within the galactic plane. Globulars all lose members when they cross

Fig. 11.3 Messier 71 is right in between γ and δ Sagittae, observed with a pair of 15×70 binoculars. The fov is 4.4° wide

the galaxy's plane. This phenomenon, called 'evaporation,' ultimately leads to total disintegration of the globular cluster. In the case of Messier 70, the globular must be in constant evaporation. It therefore will not survive as long as one of the globulars we visited in the month of June.

Locating Messier 71 is an easy task. As shown on the finder chart of Fig. 10.5, the globular resides in the small constellation Sagitta, the Arrow. A pair of 10×50 binoculars is all you need. Look for a small and vague nebula right in between γ and δ Sagittae. The sketch in Fig. 11.3 illustrates that this globular lies close to the galactic plane, hence in a very rich star field.

M71 is an eighth mag loose globular at a distance of 18,000 ly. Its visible size of 4′ corresponds with a true size of 40 ly. The brightest cluster members shine at mag 12.1. Resolution of the cluster is feasible with a 3 in. telescope although the crowd of foreground stars does not make M71 stand out very well. The globular has a mass of 40,000 solar masses, which is a rather low value for a globular cluster. For a very long time, there was doubt about the true nature of M71. Is it a dense open cluster or a loose globular? The consensus is that Messier 71 is a very loose globular. Its low mass and star count are likely a consequence of the tidal forces exerted by the galactic plane on the stars of M71.

The Cygnus Star Cloud and the Northern Coalsack

As shown in Fig. 11.1, the galactic center lies in the constellation of Sagittarius, its presence indicated by the Great Sagittarius Star Cloud. When we look in the direction of the constellation of Cygnus, our view is directed at the incurving side of our own spiral arm.

Can we perceive any features on the forwarding side of our own spiral arm with the naked eye? Yes, we can! Although our view of the galactic plane in Cygnus is partly blocked by the obscuring clouds of the Great Rift (without the Great Rift, the glowing band of the Milky Way would appear much brighter), the Milky Way from Albireo to γ Cygni is relatively free of dark interstellar matter. Here we see a bright, 20° wide oval star cloud, called the Cygnus Star Cloud, which is probably the brightest star cloud for us northern hemisphere observers. We literally look right into thousand of light-years of the cross section of the Orion Spur (see Fig. 11.4). The Cygnus Star Cloud marks the direction of rotation of the Milky Way and thus also our destination in space on our orbit around the galactic center. It is astonishing how detailed the Milky Way appears from a dark site when you take the time to make a naked-eye sketch. As such, the Cygnus Star Cloud does not remain featureless. The part from Albireo to η Cygni is rather smooth and simply bright, but the stretch from η Cygni to γ Cygni is garnished with bright stars and drenched with dark curls of the neighboring Great Rift. This section of the Cygnus Star Cloud just begs to be observed with a pair of binoculars, which we will do in the next paragraph.

A study of the Cygnus Star Cloud can't be made without noticing the protruding obscurity of the Great Rift. The Great Rift is an accumulation of giant molecular clouds that are responsible for the birth of new genera-tions of stars. These dark clouds are relatively nearby. In Aquila, the Rift is about 500 ly away and its distance increases in the direction of Cygnus. In Vulpecula, the Rift is 1,300 ly away, and in Cygnus, the Rift is at 2,300–2,600 ly from us. The distance of the Great Rift is displayed in Fig. 11.2.

The Great Rift appears to end south of Deneb (see Fig. 11.4). The Milky Way glows much brighter again east of Deneb. Here we see the second bright star cloud of Cygnus, the North America Star Cloud. A small starlit patch marks even the presence of the star field within the North America Nebula. The dark area south of Deneb, which marks the end point of the Great Rift, is called the Northern Coalsack. This area, which is visible with the naked eye at a dark location, looks extremely star poor. Even with binoculars, fewer stars can be noticed in the Northern Coalsack area than in other stretches of the Great Rift. The Northern Coalsack is believed to be a giant molecular cloud at a much closer distance of roughly 1,300 ly, thus half the distance of the Cygnus Rift.

When you have carefully observed the Cygnus Milky Way with the unaided eye, you can scan the area with your lowest power binoculars again. Try to pay attention not only to the myriads of stars, but to the subtle variations of the background glow of the Milky Way, which comes into its own with the use of your averted vision.

A naked-eye sketch of the ultimate deep-sky object, which the Milky Way is, can be very straightforward (see Fig. 2.9). It is an extremely rewarding way to discover the beautiful but subtle structure of our galaxy. You will be amazed by the power of your unaided eyes. Perhaps you can try the follow-ing approach to sketching the Milky Way. When you have drawn the stars that you can clearly see, make a second scan of the area of your interest and focus on the bright patches within the Milky Way. Once you are done with that, make a final scan while you concentrate your attention on the obscuring clouds, lanes and canals. This three-stage observation will help you to study the Milky Way in very manageable steps.

Fig. 11.4 The beautiful Cygnus Star Cloud and the Great Rift, observed with the naked eye

NGC 6871

Now that we have explored the Cygnus Star Cloud with the naked eye, it's time to bring on our binoculars. Visually the most enjoyable area is the star field NE of η Cygni, the star in the middle of the Cygnus Star Cloud (see Fig. 11.5). This area seems promising when looked at with the naked eye on a dark night (see Fig. 11.4). Indeed, the view of this stretch of the Milky Way with a pair of binoculars is far from disappointing. A labeled sketch of it is shown in Fig. 11.6.

When you place η Cygni at the SW edge of your fov, the stars 27, 28 and 29 Cygni fill the center of the field. Just south of 27 Cygni is a 20′ long star chain of mag 7 to mag 9 stars, embedded in a hazy environment. This is

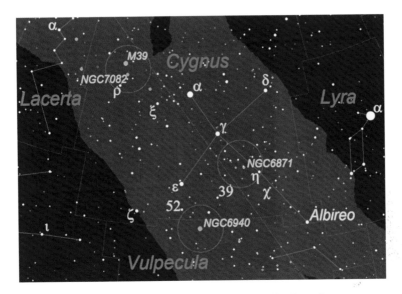

Fig. 11.5 Finder chart for the objects in Cygnus and Vulpecula

perhaps the most interesting star cluster in Cygnus, called NGC 6871. Yes, a cluster that looks like chain of stars. Notice that this fov has plenty of star chains. NGC 6871 is a just-out-of-the-cradle cluster that shines at mag 6.8 and that lies 5,100 ly away in the Orion Spur. Its true size is 30 ly.

The cluster lies at the heart of the young Cygnus OB3 Association, a 350×150 ly large section of the Orion Spur that (literally) sparkles with newly born stars, of which many are of an exotic kind. NGC 6871's lucida, for instance, is a binary system of a hot supergiant O-star and a superhot (50,000 K) Wolf-Rayet star. Wolf-Rayet stars are believed to be evolved massive O-stars that lose mass at a very high rate by means of a strong and turbulent stellar wind before they explode as supernovae. A few other Wolf-Rayet stars are found in the OB3 Association as well, but the most astonishing object in OB3 is Cygnus X-1, the first galactic x-ray source ever discovered.

Cygnus X-1 is thought to be a binary system of very massive O-stars. The most massive companion went supernova, during which its core collapsed into a 20 solar mass black hole. The remaining supergiant has about 40

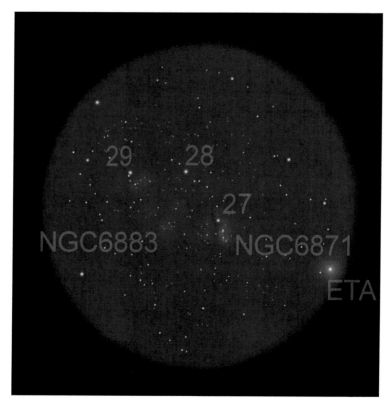

Fig. 11.6 A close-up of the Cygnus Star Cloud with a pair of 8×56 binoculars

solar masses. Both objects orbit each other in five and a half days. A black hole is an extremely dense body. The gravity on its surface is so strong that no material or even a light beam can escape from it. It is a point of no return. That's why it called a 'black' hole.

Black holes are invisible yet detectable. The black hole of Cygnus X-1 is so close to its visible companion that it pulls matter from the latter. The extracted material from the blue companion forms a disk around the black hole, from which mass spirals down into the black body. The whole process heats matter so much that it radiates its energy as x-rays. If the visible companion of Cygnus X-1 has enough mass left when it goes supernova, its core might collapse into a black hole, too. Cygnus X-1 lies about 30′ NE of η Cygni, and thus in the fov shown in Fig. 11.6. Its visible companion shines like a ninth mag star, which is visible in a small telescope.

The star field between NGC 6871 and 29 Cygni shows a few subtle brightenings in the background glow, of which NGC 6883 at 1° east of NGC 6871 is the most obvious example. NGC 6883 is a modest cluster at 4,500 ly away. Half a dozen faint stars can be seen with binoculars in this 50-million-year-old cluster. The star field at the center of Cygnus OB3 is filled with star chains. You can easily trace a large star chain from NGC 6871 over 27–28 Cygni. The field is so complex that you could forget to observe the nature of the background glow. Notice that several dark outliers of the Cygnus Rift are present in this area as well. All are best observed with averted vision.

NGC 6940

Just south of the Cygnus Rift awaits a true stellar treat. Your first impression might be that of a hook-shaped nebula, overlaid with eighth mag stars, The mag 6.3 oval brightening is in fact the open cluster NGC 6940 in Vulpeculae. A pair of 8 × 56 binoculars display a hazy patch as large as one and a half diameters of the full Moon, studded with subtle pinpricks of light. Use ε Cygni as your start point for this star hop. Then move 3° to the south towards 52 Cygni. Move another 3½° to the southwest, and NGC 6940 comes into view. When you center the cluster in your binoculars, 52 Cygni moves out of the fov.

NGC 6940 lies at a distance of 2,500 ly. This 870-million-year-old cluster is about 22 ly in size. It consists of scores of mag 9 and fainter stars. Its lucida is a mag 9 M-type red giant. Beware that the bright stars within the haze of NGC 6940 do not belong to the cluster; they are foreground or background stars. NGC 6940 appears to be at least 45′ large, though many resources give a value of 30′ for its size. The explanation could be that the eighth mag field stars look like cluster members in a pair of binoculars. The high power view in a telescope allows a better distinction between field stars and cluster members. NGC 6940 is located in the Orion Spur, as Fig. 11.2 shows. It is a pity that Charles Messier missed this beautiful cluster, as NGC 6940 could surely have been seen with the comet hunter's imperfect telescopes. This

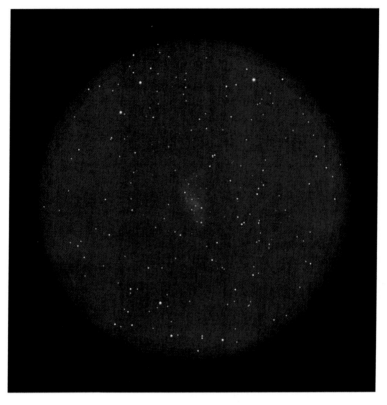

Fig. 11.7 NGC 6940, observed with a pair of 8×56 binoculars. The fov is 5.9° wide

wonderful galactic cluster would certainly have been a worthy addition to his list of comet masquerades (Fig.11.7).

Messier 39

One of the two Cygnus objects that Messier did include in his list of deep-sky objects is the mag 4.6 open cluster Messier 39, near the tail of Cygnus. For this star hop, we use Deneb as our starting point. Now move 4° and 20′ to the east and a little to the south towards fourth mag ξ Cygni. Both

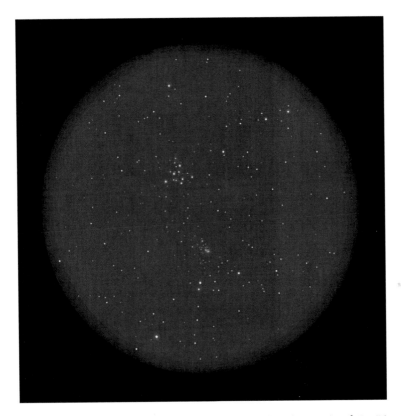

Fig. 11.8 Messier 39 and NGC 7082, observed with a pair of 8×56 binoculars

Deneb and ξ Cygni fit in the same fov of ordinary binoculars. From ξ Cygni move 5½° ENE towards fourth mag ρ Cygni. Now shift another 3° to the north until the sparse open cluster, Messier 39 appears centered in your fov. As shown in the sketch (see Fig. 11.8), both ρ Cygni and M39 fit nicely in the same fov. If your sky is dark enough, you can try to make this star hop without optical aid. Messier 39 will show up as a small patch in the rich Milky Way clouds at 9° ENE of Deneb.

In a pair of 8×56 binoculars, Messier 39 is shown to consist of a dozen seventh–ninth mag stars in a triangular arrangement. This open cluster is physically one of the smallest Messier clusters. That it appears large is due to its close distance to us. M39's angular size of 30′ corresponds with a true size of 9 ly at a distance of 1,000 ly. The cluster is best appreciated with ordinary binoculars, because it is too scattered to be observed with a telescope. About 30 stars are believed to belong to M39. All of them are main sequence stars. The lucida of M39 is a seventh mag B-star. The other bright stars are mainly A-subgiants. The cluster is 240–480 million years old. Although sparse, the total luminosity of M39 is about 830 Suns. A star like our Sun would be invisible in our binoculars if placed near M39 because it would shine at mag 12.

Messier 39 is not the only object in our fov. About 1½° south of M39 emerges a tight fuzzy knot of minute stars, known as NGC 7082, a seventh mag open cluster. Its shape reminds one of a comet when you use averted vision. A pair of 8 × 56 binoculars resolve about a dozen stars.

NGC 7082 and M39 are all but neighbors (see Fig. 11.2). Although M39 lies closer to the inside of the Orion Spur, NGC 7082 is located almost five times further away, near the outside of the Orion Spur. NGC 7082 is a 80-million-year young cluster at a distance of 4,700 ly. Its angular size of 25′ matches up with a true dimension of 35 ly. From our viewpoint, M39 appears brighter than NGC 7028. In absolute numbers however NGC 7082 shines two magnitudes brighter than M39. But the image that we see of NGC 7082 is about 3,700 years older than the one of Messier 39. Our binoculars have become a small time machine. When we look at NGC 7082, we look further back in time than when we look at Messier 39.

Fomalhaut

The southern autumn sky (Fig. 1.7) is a rather desolate area. It is occupied by dim water-related constellations such as Aquarius, Pisces, Cetus and Piscis Australis (often called Piscis Austrinus). The square of Pegasus is the most obvious and easily recognizable star pattern. Not many bright stars or asterisms are found south of Pegasus. The one exception is the first magnitude star with the exotic name Fomalhaut in the deep southern constellation of Piscis Australis, the Southern Fish. Fomalhaut comes from the Arabic 'fum al-hawt' and means 'mouth of the fish.' If there is one star that should be associated with autumn then it is Fomalhaut. Its low appearance above the southern horizon for us northern hemisphere observers, together with the star-barren constellations above it, turn

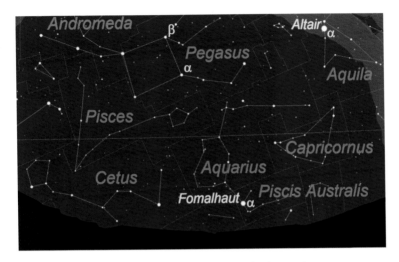

Fig. 11.9 Finder chart for Fomalhaut, as it rises in the southeast on autumn nights

Fomalhaut into a isolated and lonely beacon. Observers north of 60° latitude never see Fomalhaut rise. Even for midnorthern observers, Fomalhaut stays low above the southern horizon. You will need a clear view to the south to see the lonely autumn star.

Locating Fomalhaut is fairly easy. Just draw a line from the westernmost stars of the square of Pegasus, β and α Pegasi, to the south for 45° (a little more than two hands spread wide open) and you arrive at Fomalhaut (see Fig. 11.9). Wait until late into the evening, when the square of Pegasus is high enough in the sky for Fomalhaut to rise above the horizon.

Of course, Fomalhaut's loneliness exists only in our imagination. Southern hemisphere observers around 35° latitude see Fomalhaut as their zenith star. In a way, Fomalhaut is their counterpart for our (northern resident) zenith star, Vega (see the previous chapter). In space, Fomalhaut isn't lonely either. As shown in Fig. 4.2, it lies relatively below the galactic plane, though still close to our Sun.

Fomalhaut shares a few more characteristics with Vega. They both lie at a distance of 25 ly, and they both are A-type stars. Fomalhaut is a little cooler and smaller than Vega. Its mass is twice the Sun's, and its diameter is 1.8

Fig. 11.10 Fomalhaut, observed with a pair of 8 × 40 binoculars

times that of the Sun. Fomalhaut has a surface temperature of 8,500 K. Infrared observations have indicated that Fomalhaut is surrounded by a large disk of dust. Further investigations have led to the first ever directly imaged planet. It orbits Fomalhaut from within the disk of dust. There are possibly more planets around Fomalhaut, but only one has been discovered so far. Whether life like we know it here on Earth can develop in the Fomalhaut system is doubtful. The star is about 200 million years old and has an estimated lifespan of only 1 billion years. Life on Earth needed about 3.5

billion years to develop to its current state. It was only possible because our Sun has an estimated lifetime of approximately 10 billion years (Fig. 11.10).

Although Fomalhaut is a true first magnitude star and shines with the light of a pure white star, mid-northern observer see it more like a slightly reddened mag 1.5 or mag 2 star, because of the dimming effect of the thicker atmosphere we look through near the horizon.

CHAPTER 12

OCTOBER

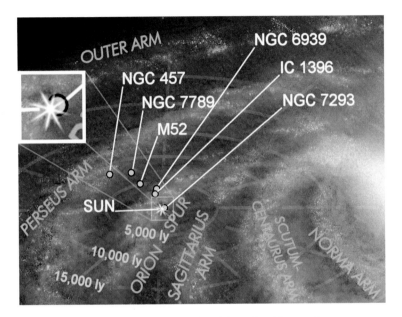

Fig. 12.1 The location of the deep-sky objects for this month

During October nights, the winged horse Pegasus flies high through the sky. Its westward motion is followed in a loyal manner by the lonely autumn star, Fomalhaut. Our night-time window is slowly turning towards the rim of the Milky Way where new constellations with exciting stellar treasures await our visit (Fig. 12.1).

NGC 7293

Before we turn our binoculars towards the rim of our galaxy, let us pay a last visit to the inner edge of the Orion Spur, in the direction of the dim constellation of Aquarius. Here we find one of the most amazing nebulae of the whole

R. De Laet, *The Casual Sky Observer's Guide*, Astronomer's Pocket Field Guide,
DOI 10.1007/978-1-4614-0595-5_12, © Springer Science+Business Media, LLC 2012

sky, an object that many astronomers have failed to see simply because it looks brighter in a modern pair of binoculars than in a classic telescope. The object of our interest is NGC 7293, nicknamed the Helix Nebula.

NGC 7293 is the largest and brightest planetary nebula. Its size of 12' (a third of the disk of the full Moon!) and its total brightness of mag 7.3 could make you think that the Helix Nebula is an easy object to observe. Well, the contrary is true. The nebula's surface brightness is rather low. If you use a telescope with a high magnification, the planetary's light is smeared out too much for your eyes to see it against the background glow. The problem gets even worse if light pollution comes into play. A pair of binoculars, on the other hand, has the ability to take in a lot of sky and to focus the planetary's figure into a smaller and thus brighter image. As such the contrasts between the nebula and the sky remain sufficiently high for our eyes to see it (Fig. 12.2).

NGC 7293 lies at a distance of 700 ly. Its true size is about 4 ly, or half the distance from the Sun to Sirius. In its center glows the remnant of a red giant that has shed its outer layers. These now form the expanding nebula.

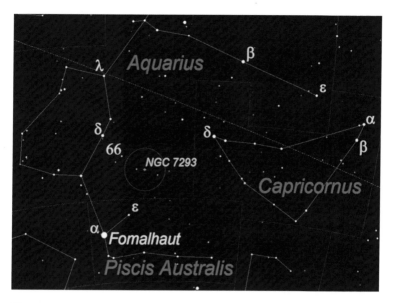

Fig. 12.2 Finder chart for the Helix Nebula

The central star is a 13th magnitude hot white dwarf with a surface temperature of 110,000 K. Thanks to the energetic ultraviolet radiation of the white dwarf its expelled outer layers light up.

The nebula is believed to have the shape of a barrel and is therefore nicknamed the Helix or Helical Nebula. We happen to look into the open end of the barrel. NGC 7293 is about 10,000 years old. Because its expansion never ceases, it will ultimately grow so large that it will become too dispersed to be noticed at all. This is the reason why we see so few planetaries. They represent an extremely short phase of a star in its death throes.

Unfortunately the Helix Nebula lies in a star-barren field. The best possible star hop starts at the third mag δ Aquarii. From here, move 4° to the SW to fifth mag 66 Aquarii. Finally move just less than another 4° to the WSW. The Helix now lies in the middle of the fov. Make use of your averted vision and look for a little disk that is barely any brighter than the background. Figure 12.3 shows what you might expect to see.

Once you have found this fine planetary, take all the time your eyes need to form a good impression of its subtle light. From a dark and moonless site, the Helix is a great binocular target. A pair of 8×56 binoculars display NGC 7293 as a round and gossamer smudge of light. Two faint stars seem to border the western rim of the Helix. The planetary's disk shows no structure.

With 15×70 binoculars, the Helix appears larger and slightly brighter against a darker background. Fainter stars surround the nebula. The center of NGC 7293 looks dimmer than its perimeter. The northern rim of the Helix is the brightest feature of the disk, and this strengthens the impression that the planetary is hollow. If you want to observe this nebula with a telescope, use your lowest magnification at first and a UHC filter. The nebula is visible with powers ranging from 16× up to 42×, while it simply vanishes at higher powers. When you remove the UHC filter from the eyepiece, the nebula disappears as well. You can count yourself part of a select club of observers if you have managed to see the Helix, because many backyard astronomers fail.

The progenitor of NGC 7293 no longer exists. The star has thrown off its outer layers into space. Its lifeless but still-hot core will very slowly cool down. Maybe the star had a planet system like our Solar System.

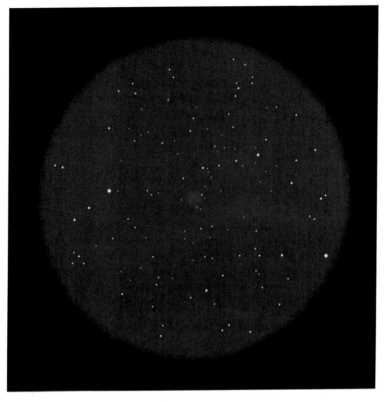

Fig. 12.3 NGC 7293, the Helix Nebula, observed with a pair of 15×70 binoculars. The fov is 4.4°

Perhaps there was life on one of its planets. When this star died, worlds would have been destroyed, its ashes ejected into the interstellar medium, of which new stars would be born. This is a major lesson for us to learn. It is the destiny of many stars, including our Sun, to end their lifecycle with the development of a planetary nebula. We take our Sun for granted. It is always there. Our Sun has given life to Earth, but it will also take life back when the time comes (in about 5 billion years). It is neither our concern, nor our children's. Will humankind survive the final destiny of our Solar System, or will our species return to the stellar ashes from which it was created in the first place?

NGC 6939 and 6946

When we follow the glowing band of the Milky Way further east of Cygnus, we look through the Orion Spur towards the outer rim of our galaxy. Here the star clouds are not as dense as in the direction of the galactic center, but the autumn constellations near the galactic plane offer some very fascinating sights for binocular and small 'scope observers. Our next destination lies in the southwestern part of Cepheus, a doghouse-shaped constellation northeast of Cygnus (see Fig. 1.8). Here we find an interesting pair of rather faint deep-sky objects about 40′ apart.

For this star hop, we start from second mag α Cephei and move 4° west to third mag η Cephei. Now we move 2° SW. In the middle of our eyepieces we can see the gossamer light of two hazy smudges, called NGC 6939 and NGC 6946 (Fig. 12.4).

The sketch in shows Fig. 12.5 you what to look for. At first glance both objects appear similar in size and brightness. It takes time to notice that the northern object, NGC 6939, is a tad brighter than its southern sibling.

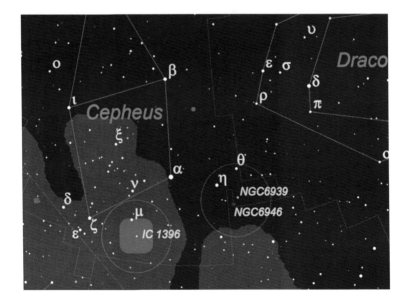

Fig. 12.4 Finder chart for the Cepheus objects

Fig. 12.5 NGC 6939 and 6946, observed with a pair of 15×70 binoculars. The fov is 4.4° across

Notwithstanding their similar appearance, both objects have nothing in common whatsoever. They don't even share the same location in space.

NGC 6939 is a mag 7.8 open cluster at 4,100 ly away. Its age is about 1.6 billion years. Because of its age, the loose cluster managed to drift about 10° away from the galactic plane. The galactic cluster is located near the outer rim of the Orion Spur. A pair of 15×70 binoculars shows a 7′ wide haze with a slightly brighter core. A few faint stars are embedded in the amorphous glow.

NGC 6946, on the other hand, is a loose spiral galaxy 5,000 times more distant than NGC 6939. Its integrated brightness of mag 8.8 is spread over an area as large as 11'. The galaxy's true spread is about 60,000 ly at a distance of 20 million ly. That NGC 6946 lies close to the galactic plane is illustrated by its location in Fig. 5.2. The galaxy is legendary for its nine supernovae that have been discovered since 1917. In 15×70 binoculars, NGC 6946 displays a soft halo that gradually brightens towards its core.

Both objects are greatly dimmed by the interstellar dust in the direction of Cepheus. It is impossible to separate the galaxy from the galactic cluster with a small pair of binoculars. Imagine how often the classic astronomers went astray when qualifying the true nature of a deep-sky object. The galaxy does not appear to be a full magnitude dimmer than its northern cousin is, perhaps because we can only see the bright core, which has the highest surface brightness of the whole object. The overall view of Fig. 12.5 gives us a real sense of depth in space. The field stars are relatively nearby foreground objects at maybe 100 to 1,000 ly away.

NGC 6939 marks the edge of our local spiral arm at a distance of 4,000 ly. NGC 6946, on the other hand, is a foreign object that lies far beyond the border of our galaxy. It is an isolated island of billions of stars at 20 million ly away.

IC 1396 and Herschel's Garnet Star

Cepheus, the doghouse-shaped constellation, harbors many celestial treasures. The reason is obvious. On a dark and moonless night, the bright Milky Way can be seen northeast of Cygnus, protruding far into the box-shaped part of Cepheus. Here resides one of the largest stellar nurseries of the Orion Spur, the emission nebula called IC 1396. Its presence is even visible with 8×40 and 10×50 binoculars. IC 1396 is a true tracer of the Orion Spur. It forms the heart of the Cepheus OB2 Association, a very young stellar association whose dense dust clouds are responsible for the absence of the soft Milky Way background glow in southern Cepheus. Cepheus OB2 is believed to be 2,500 ly away.

Locating IC 1396 is easy, as it lies at the entrance to the doghouse. The position of IC 1396 is marked by its own guard of honor, namely μ Cephei – the Garnet Star (see Fig. 12.4). You cannot miss fourth mag μ Cephei, nor its deep-orange color cast, for which the great eighteenth century observer William Herschel called it 'the Garnet Star.' Mu (μ) Cephei is an over-luminous red M-type supergiant. It is, with 1,600 solar radii, one of the largest stars in our galaxy. If it were placed in the middle of our Solar System, its cool surface of 3,700 K would reach somewhere between the orbits of Jupiter and Saturn.

Mu (μ) Cephei was likely born only a few million years ago in the Cepheus OB 2 Association. Its initial mass was probably over 40 solar masses. The star quickly exhausted its core hydrogen supply and is now nearing its death as a huge red star. The star's deep-orange color is the result not only of its low surface temperature but also by a shell of ejected material that surrounds the star and by the interstellar dust in our line of sight. Mu (μ) Cephei is a variable star. Its brightness varies from mag 3.6 to mag 5 with a period of 900 days. The fate of μ Cephei is to go supernova in the near (relatively speaking) future. The shockwave following the explosion will probably trigger the birth of new stars in the nearby dust clouds of Cepheus OB2.

For the time being, μ Cephei leads the way to IC 1396, as shown in Fig. 12.6. A pair of binoculars will reveal both the Garnet Star and IC 1396 in the same field of view. The tenuous glow of the emission nebula needs a lot of dark background to stand out. Because IC 1396 spans two and a half degrees, a telescopic field of view might be too narrow to show the emission nebula clearly against the backdrop.

At first, IC 1396 seems to be an ethereal amorphous glow, whose presence is only confirmed with averted vision. In the middle of the nebula shines the fifth mag multiple star Struve 2816. The primary of Struve 2816, a very hot O-star, is believed to be responsible for the glow of IC 1396. Its strong UV radiation lights up the whole IC 1396, an area as large as 100×120 ly! Struve 2816 is also the brightest star of the loose cluster (or should we rather say star condensation) within IC 1396, called Trumpler 37. IC 1396 comes more into its own if you concentrate on its dark channels and voids, of which there are plenty to discover on a dark and moonless night.

This star birth region is extremely active. There are multiple clouds slowly condensing into new stars. Once formed, their strong stellar winds blow large cavities in the dense dust clouds. The most massive stars soon grow large and red before they explode as supernovae. With both Struve 2816 and μ Cephei side by side, we have youth and age gathered in our field of view. Such a young and active stellar association must be a very

Fig. 12.6 μ Cephei and IC 1396, observed with a 4-in. refractor at 20×. The fov measures 2.5° across

uncomfortable place for inhabited planets such as Earth. Fortunately we can witness the stellar evolution of Cepheus OB2, with its hazardous environment, from a safe distance of 2,500 ly.

Messier 52

In the direction of Cepheus, the Milky Way is partly obscured by the dense dust clouds of the Cepheus OB Association. The adjacent constellation of Cassiopeia, on the other hand, is much less affected by interstellar dust.

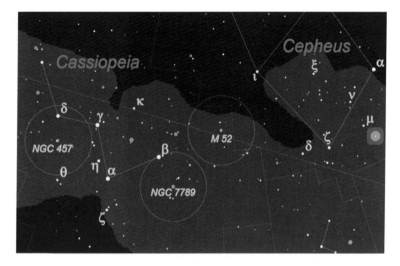

Fig. 12.7 Finder chart for the Cassiopeia objects

It offers us a clear and deep view, the Cassiopeia Window, in the direction of our galaxy's rim. No wonder Cassiopeia is the favorite hunting ground for autumn observers.

Our next target along the galactic equator is the bright but compact open cluster, Messier 52. This cluster lies almost right on the galactic plane (see Fig. 12.7). We can use α and β Cassiopeiae to find M52. When we draw an imaginary line through them towards Cepheus for another 6° (about one binocular field), we arrive at Messier 52.

Messier 52 appears like a small fuzzy spot in an ordinary pair of binoculars. With a magnitude of 6.9 it is a fairly bright but tight knot in 15×70 binoculars. You can just about resolve a dozen stars with averted vision in this 12′ cluster. It's lucida, a yellow G-type giant, stands out prominently at mag 8.2 near the southwestern edge of the cluster. The other cluster members are nearing mag 11 or dimmer (Fig. 12.8).

Messier 52 is, after M11, one of the densest open clusters. That is why it looks more impressive at the higher power of a small telescope. The cluster is believed to consist of more than 6,000 stars. Its absolute brightness

Fig. 12.8 Messier 52, observed with a pair of 15×70 binoculars. The fov is 4.4° wide

equals about 20,000 times that of the Sun's. Messier 52 lies in the gap between the Orion Spur and the Perseus Arm, at a distance of 4,600 ly. It has a true size of 22 ly. Its age is estimated at 69 million years.

NGC 7789

Perhaps the most impressive Cassiopeia object is the large galactic cluster NGC 7789, at 3° SSW of β Cassiopeiae. This mag 6.7 gem is almost as large as the disk of a full Moon. Like M52, NGC 7789 is an 'inter-arm gap' object that lies at 7.600 ly from us. Its true size is roughly 50 ly. This 1.7-billion-year-old cluster shines with an absolute brightness of about 12,000 Suns. Considering its age, NGC 7789 is quite a brilliant object.

Being the most attractive Cassiopeia object, NGC 7789 merits all the close attention you can pay to it. Its western half appears the brightest and exhibits the sharpest boundary. The cluster's eastern half is rather dim and its edge is ill defined, like the head of a comet. If you take the time and are patient, the cluster will seem to grow larger in the eyepiece. Countless tiny stars jump in and out of view with averted vision, an immersive experience behind the eyepiece that no photograph can mimic (Fig. 12.9).

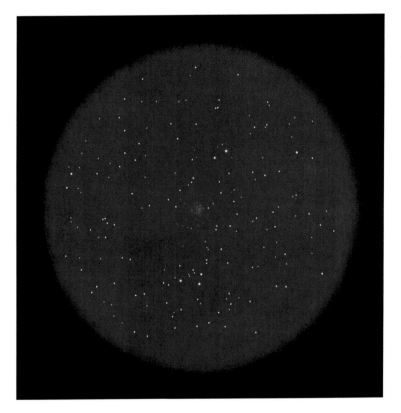

Fig. 12.9 NGC 7789, observed with a pair of 15 × 70 binoculars

NGC 457

Maybe the most inspiring object of the autumn sky is NGC 457 at 2° SSE of δ Cassiopeiae. NGC 457 is often called the ET cluster, for the object's resemblance to the alien visitor in Steven Spielberg's movie *The Extra-Terrestrial.* Indeed, the telescopic appearance of the 13' large galactic cluster at 20× is nothing short of spectacular. The cluster's stars mimic ET's body perfectly, including his large eyes (φ Cas and its companion), waving arms and long feet. Of course, the cluster was known long before that. Its former nickname was the Phi Cas Cluster, referring to NGC 457's mag 5 brightest star, φ Cassiopeiae. However the Phi Cas Cluster was not a popular object among stargazers until ET hit the big screen. Since then NGC 457 has grown more and more popular.

Still, through binoculars at low power, NGC 457 looks more like the fuselage of a fighter jet with its afterburners, represented by φ Cas and its seventh mag companion. Only a handful of stars can be resolved in the central bar of the cluster. A pair of 15 × 70 binoculars show a better version of ET, as power and aperture bring the fainter stars into view. Although NGC 457 is a very interesting binocular object, the Extra-Terrestrial gives its best performance in a small telescope at medium magnification (Fig. 12.10).

The true nature of NGC 457 is still not clear. It is believed to lie at 7,900 ly from us in the Perseus Arm of our galaxy. The cluster is about 24 million years old. φ Cas was long thought of as the lucida of NGC 457. But a yellow F-star of fifth magnitude must be quite massive in order to be so bright from such a distance. Massive stars die rather young, and they certainly don't exist in 24-million-year-old clusters. Thus φ Cas and its companion are probably foreground stars at a distance of 4,600 ly. If this is true, then NGC 457's lucida is a ninth magnitude M-type red supergiant in the 'body' of the cluster. Without φ Cas in its membership, the cluster's brightness is rated at mag 6.4.

If our assumptions are correct, then φ Cas and its companion are local residents of our own spiral arm. They just happen to lie in the same line of sight as an 'alien' cluster, which is located in the Perseus Spiral Arm.

Fig. 12.10 NGC 457, observed with a pair of 15×70 binoculars

Many observing guides are not clear about the matter, either. They list the NGC 467's apparent brightness at about mag 6.4 but meanwhile assume that fifth mag φ Cas is a member! The companion of φ Cas is a blue B-star. See if you can detect the color difference between yellow φ Cas, its seventh mag blue companion, and eventually the ninth mag lucida of NGC 457.

CHAPTER 13

NOVEMBER

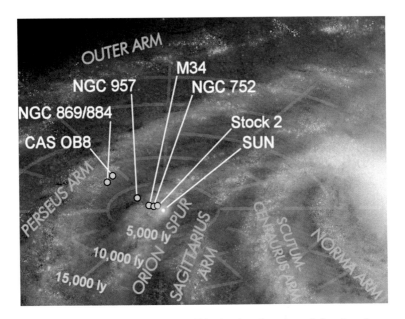

Fig. 13.1 The November objects all lie in the direction of the rim of our galaxy

This month, the glowing band of the Milky Way in the direction of the beautiful constellations of Cassiopeia and Perseus passes high overhead. It offers us an excellent opportunity to go deep-sky hunting in the Cassiopeia Window of the Milky Way. This window is a relatively dust-free stretch of our local spiral arm with a clear view towards the next exterior spiral arm of our galaxy; the Perseus Arm (Fig. 13.1).

November evenings also feature the famous Andromeda constellation, which harbors the magnificent Andromeda Galaxy, M31. This and many other delightful binocular targets await us this month.

R. De Laet, *The Casual Sky Observer's Guide*, Astronomer's Pocket Field Guide, DOI 10.1007/978-1-4614-0595-5_13, © Springer Science+Business Media, LLC 2012

Messier 103 and the Cassiopeia OB8 Association

Our November tour starts with a Messier object, M103. It is a bit puzzling why Messier included this cluster in his list, while Cassiopeia has so many brighter comet-like objects to mention. Charles Messier added M103 because of a report that he received from his colleague Pierre Méchain. He probably did not observe M103 himself.

Messier 103 is visible with a small pair of binoculars as a little knot of two or three 7th to 8th magnitude stars, 1° ENE of third magnitude δ Cassiopeiae. As such, the cluster does not impress. What makes M103 worth a look is its company. A total of four galactic clusters await our attention. With the use of Fig. 13.2 and the sketch in Fig. 13.3, you can find the location of Messier 103 and its related clusters.

Messier 103 is one of the remotest galactic clusters of Messier's list. It lies at 7,200 ly away, in the Perseus Arm of our galaxy. Messier 103 is believed to be part of the Cassiopeia OB8 Association, which is marked on Fig. 13.1. M103 has an apparent brightness of mag 7 and measures 6' across. Messier 103's

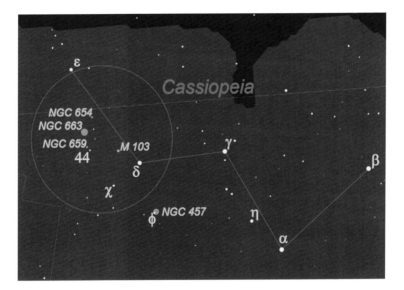

Fig. 13.2 Finder chart for the Cassiopeia objects

Fig. 13.3 Messier 103, NGC 654, NGC 659 and NGC 663, observed with a pair of 15×70 binoculars. The third magnitude δ Cassiopeiae marks the right edge of the sketch. The fov measures 4.4° across

true dimension is about 16 ly. The cluster's brightest star is a seventh magnitude blue B5-type supergiant near the northern edge of the cluster. It shines with the equivalent luminosity of 40,000 Suns. The second brightest member of M103 is an eighth mag red M-type supergiant. Both stars are the only members that small binoculars can reveal. A small telescope will show half a dozen stars more, in a triangular shape. The cluster's age is estimated at 24 million years, about the same age as the Cas OB8 association.

A visually more interesting object is NGC 663, at one and a half degrees away from M103, on the same line with δ Cas. This galactic cluster looks like a 15' large field condensation of minute stars. Its integrated brightness is also mag 7. NGC 663 lies 6,400 ly away and is 28 ly large. It is also part of the Cas OB8 Association but originated only 14 million years ago.

With care, two other galactic clusters can be seen near NGC 663. At about 30' SSW of NGC 663 lies the eighth mag open cluster NGC 659. It looks like a fuzzy spot that is 5' large. NGC 659 is 6,300 ly away. Another open cluster is NGC 654, at 40' NNW of NGC 663. Its 5' large glow has the brightness of mag 6.5. The cluster is 6,700 ly away from us.

All four clusters, which likely belong to the Cassiopeia OB8 Association, can be seen in the same field of view with a pair of binoculars. The fact that our binoculars are able to see so far is due to the transparency of the Cassiopeia Window in the autumn Milky Way. The luminosity of the above mentioned clusters is only slightly tempered by the nearby interstellar dust clouds of the galactic plane (Fig. 13.3).

The Double Cluster of Perseus

Much to the delight of the binocular observer, the Cassiopeia Window stretches deep into the Perseus constellation as well. The splendor of the Milky Way in Perseus is even obvious to naked-eye observers. Perseus offers us a mix of naked-eye clusters from both within our local spiral arm as well as from the Perseus Arm.

The most spectacular open cluster is undoubtedly the Double Cluster of Perseus. The presence of the two bright star clusters in Perseus did not go unnoticed to our ancestors. The Greek astronomer Hipparchos was the first to catalog them in 130 B.C. On a dark night, we can repeat Hipparchos' naked-eye observation. With both Perseus and Cassiopeia high in the sky, draw a line from δ Cassiopeia to γ Persei. Between these stars, there is a brighter area in the Milky Way. If you remain patient, you can distinguish two adjacent bright patches, each nearly as large as the disk of the full Moon. These magnificent clusters, cataloged as NGC 884 and NGC 869, are true tracers of the remote Perseus Arm. Their light, which is dimmed about 1.5 magnitudes by interstellar dust, has traveled for 7,000 years to reach your eyes! Most observing books mention that the Double Cluster can only

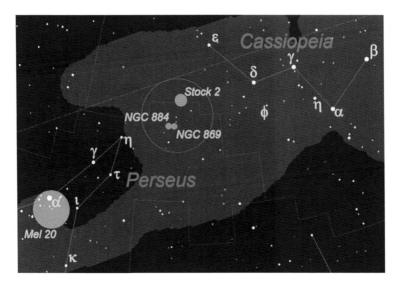

Fig. 13.4 Finder chart for the Double Cluster

be seen with the naked eye as one single patch of light; this is inaccurate when observing from a good sky (Fig. 13.4).

Observing the Double Cluster of Perseus with a pair of binoculars is perhaps the most enjoyable pursuit of November nights. Both clusters look like bright glittering jewel boxes in a star-studded Milky Way field. In each cluster, the scintillating stars seem to be embedded in diamond dust, an effect caused by numerous unresolved stars. The cherry on the cake for binocular observers is the presence of a third cluster, called Stock 2, with its location marked by a star chain north of the Double Cluster.

While you will enjoy the breathtaking view of these three marvelous stellar gems with 8 × 56 binoculars, a fourth object might emerge out of the backdrop. It is the subtle glow of NGC 957, one and a half degrees east of NGC 884. Instead of describing the appearance of this magnificent foursome of clusters, check out the sketch in Fig. 13.5. Each of these objects can be observed in greater detail with a telescope but this will happen at the expense of losing sight of the other clusters.

Fig. 13.5 The Double Cluster and Stock 2, observed with a pair of 8 × 56 binoculars. The fov measures 5.9° across

The Double Cluster seems to be a pair of related objects. However, distance measurements are not accurate enough to tell for sure. Both clusters belong to the same association but did not originate simultaneously. NGC 869 is populated with more evolved stars while NGC 884 has much younger members. According to recent studies, the clusters are separated from each other by roughly 800 ly. If that is correct, then the clusters are not gravitationally bound to each other.

NGC 869 is a 60 ly wide cluster at a distance of 6,800 ly. Its integrated brightness is of mag 5.3. The cluster's age is 19 million years. The mag 5.1

NGC 884 lies at a distance of 7,600 ly and measures 65 ly across. NGC 884 is a bit younger with an age of 13 million years. The luminosity of each cluster could reach about 160,000 Suns. In a pair of 8 × 56 binoculars, both clusters each reveal a dozen stars. The Double Cluster is at its best in a small telescope at a magnification of 25 times. Then both groups fill brilliantly the whole the field of view. It is a dazzling sight in a 4-in. 'scope, not only because of the spectacular radiance of the resolved star chains but also for the reason that a subtle network of dark channels seems to be holding all the stars in place as well.

How bright (or faint) would a star like our Sun shine when placed in the Double Cluster? It would not show up in even the largest amateur telescopes, for it would have a magnitude greater than 16. The reason that we can actually see individual stars in such remote clusters is because they are luminous giants.

A little more than 2° north of the Double Cluster you'll find the hazy glow of another galactic cluster, called Stock 2 (St 2). Stock 2 spans a whole degree of sky in a pair of 8 × 56 binoculars and is bespangled with dozens of minute stars. If St 2 was not so close to the Double Cluster, it would be credited with much more attention and rightly so. On the other hand, the Double Cluster is the ideal signpost to locate St 2 because of the star chain that leads the way to the huge cluster.

Stock 2 bears the charming nickname the Muscle Man, for its prominent X-shaped star chains of mag 9 stars. When you gaze long enough, the stick-figure will show up. Stock 2 appears large because it is located within our own spiral arm at a close distance of 1,200 ly. Its true size is 20 ly. The galactic cluster has already burned out its most massive stars and is believed to be about 150 million years old. You can count about 40 stars in St 2 with your 8 × 56 binoculars. A small telescope will double that number. If you observe from a dark site, you can try to detect St 2 with the naked eye, as the cluster's integrated brightness is calculated at mag 4.4. However, don't underestimate this exercise. Even though Stock 2 is about a whole magnitude brighter than NGC 869, its light is also spread over an area roughly four times as large.

The subtle shimmer, one and a half degrees east of NGC 884, comes from NGC 957, a mag 7.6 galactic cluster at 3,300 ly away near the outer edge of the Orion Spur. Only a few stars can be resolved in 8 × 56 binoculars.

The gathering of these four deep-sky objects that an ordinary pair of binoculars can display represents one of the most magnificent sights an amateur observer can dream of. Both Stock 2 and NGC 957 are evolved foreground objects, located within our local spiral arm. They happen to lie in the same line of sight with the splendid Double Cluster, which lies beyond in the next spiral arm, the Perseus Arm.

Messier 34

Another highlight of the Perseus Milky Way is the bright naked-eye cluster, Messier 34. On a dark and moonless night, it can be seen as a small hazy cloud between β Persei and γ Andromeda (see Fig. 13.6). This mag 5.2 galactic cluster lies 1,600 ly away and is thus located in our local spiral arm. Its apparent size of 35′ corresponds with a true dimension of 17 ly. Messier 34 is believed to be 260 million years old. Its lucida is a mag 7.3 B8 star with a true luminosity of 275 Suns.

The cluster is a delightful object for binocular observers. A pair of 8 × 56 binoculars show a bright stellar condensation of about a dozen stars.

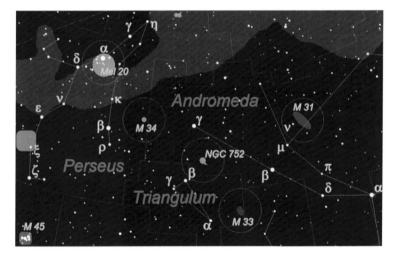

Fig. 13.6 Finder chart for the November objects

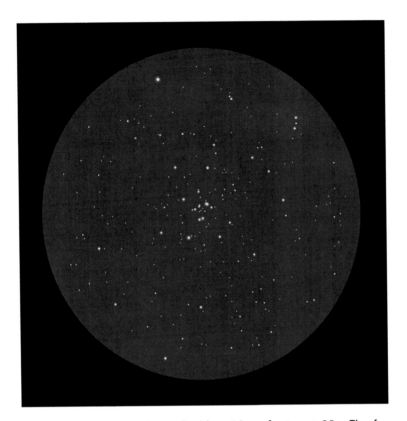

Fig. 13.7 Messier 34, observed with a 4-in. refractor at 25×. The fov measures 2° across

The view in a small refractor at low power is enchanting. About 30 stars up to mag 12 are loosely gathered in front of a rich star field. The sketch in Fig. 13.7 shows how loose the cluster actually is, when observed at 25 times magnification.

Messier 31

In the pre-telescopic era, astronomers had already discovered a large nebula in the constellation of Andromeda. It was called the Great Andromeda Nebula. The first telescopic observations date from the seventeenth century. The Andromeda Nebula then reminded one of a candle flame seen through a horn. Charles Messier included the Andromeda Nebula as the 31st entry in his catalog. He must have been enchanted by the nebula, as he took the time to make a sketch of it. He observed the object with several telescopes but could not detect any stars in M31.

The true nature of Messier 31 was beyond the imagination of the early astronomers. They assumed that the nebula was a nearby object that belonged to our Milky Way. By the end of the nineteenth century, the first photographs proved that M31 possesses a spiral structure. At that time, such objects were believed to be nascent solar systems within the Milky Way. Finally, in 1923, the American astronomer Edwin Hubble determined the true distance of Messier 31, which today is measured at 2.6 million light years!

Hubble's discovery marked a milestone in the understanding of the scale of our universe. Until then, the universe was believed to be completely occupied by the Milky Way. Hubble proved that Messier 31 was a separate island of stars that was too far away to belong to the Milky Way. It dawned upon astronomers that the universe was much larger than any of us could imagine.

Messier 31 appears to us in the northern hemisphere as the largest and brightest galaxy. It is therefore visually the most interesting galaxy to observe. With the naked eye, it is visible as a 2° oval cloud that shines with the brightness of a mag 3.5 star. When you try to locate M31, start from α Andromedae. Then move over δ to β Andromedae. From β it is a small hop over μ to ν Andromedae. M31 lies 1°21′ West of mag 4.5 ν Andromedae.

Two and a half million light-years is an incredible distance. But on a moonless night, it is not difficult to see the Andromeda Galaxy with the naked eye. Messier 31 is one of the remotest objects that can be seen without optical aid. When we look at the Andromeda Galaxy, we travel back in time by 2.6 million years. Consider that the light that our eyes nowadays collect from M31 departed from the galaxy during the dawn of the Stone Age on Earth, when the first humanlike species populated Earth. Our Solar System has completed 12 orbits around the core of our galaxy since. Another bracing contemplation is that M31 and our Milky Way are gravitationally bound to each other. The mutual attraction force is so strong that both galaxies will collide with each other in a few billion years. When you observe M31, think of the fact that we are on an unavoidable collision course with the Andromeda Galaxy. Collisions or mergers among galaxies happen frequently. Galaxies such as M31 and the Milky Way are probably the result of former collisions of smaller galaxies.

Messier 31 is an impressive object for binocular observers, as it is the archetype of all the galaxies. It prominently shows the three well defined zones commonly present in the majority of the galaxies: halo, core and nucleus. Because M31 is so large and bright, it is the best object for novice deep-sky observers to hone their observing skills. But don't let our enthusiastic descriptions rise your expectations too high.

The excitement of finding the largest galaxy in the sky can soon be replaced by serious displeasure. You might expect to see the galaxy as it appears on a stunning photograph. But Messier 31 may look only like a dim candle seen through frosted glass. You might not be able to make out any details at all. Observing experience will allow you to see more details in M31 with a pair of binoculars than inexperienced eyes could make out in a large telescope. So be warned – deep-sky observing is an acquired skill. Your first visual observation of Messier 31 will most likely not reveal all the details that are present in the sketch in Fig. 13.8. Only with long and patient observation can such a sketch be completed. And even then, no sketch can stand the comparison with the well known colorful pictures of the Andromeda Galaxy.

Cameras are much better at revealing crisp details of dim objects than our eyes. But our eyes are better suited to capture the high dynamic range of the light emitted by such galaxies. With the combined use of direct and averted vision, our eyes are able to reveal the galaxy's bright but tight nucleus, centered in the broad bulge, simultaneously with the largely extending although delicately fading oval halo. This is the domain where our eyes excel. No single picture can imitate such a live experience at the eyepiece. On the darkest of nights, M31 even shows two dust belts. The most prominent one touches the northwestern edge of the galaxy's bulge. It also indicates which side of the galaxy is turned towards us. The second dust belt runs parallel with the first one. Its presence is revealed by the sharp border of the galaxy's halo outside of the first dust belt. These dust belts are in fact similar to the Great Rift of our own Milky Way.

Fig. 13.8 The Andromeda Galaxy with its satellites M32 and M110. ν Andromedae is the bright star left of M31. The sketch is made with a pair of 15×70 binoculars. The fov measures 4.4° across

Now take a good look at the elongated shape of the galaxy's disk of stars, also called the halo. On its extremities, the halo seems to dissolve imperceptibly into the background. There is no clear border. How far the halo extends, depends on the quality of the sky but also on your visual perception. Studying the extremities of M31's halo is an excellent way of training your averted vision. The more you study the galaxy's captivating appearance, the better you will be able to observe fainter deep-sky objects as well. Some nebulae or galaxies appear no brighter than the Andromeda Galaxy's outer halo.

The Andromeda Galaxy is believed to be about 157,000 ly large. It is the largest galaxy of the Local Group. The Local Group consists of about 30 galaxies, of which M31 and the Milky Way are the largest systems. M31 is the absolute brightest galaxy of the Local Group, even when we see it on edge. Our Milky Way would shine at mag 3.9 when placed next to M31. It would also measure barely two-thirds of M31's size. The positions of all the galaxies that are featured in this book are shown in Fig. 5.2.

Being the largest galaxy of the Local Group, it is no surprise that Messier 31 has caught about 12 smaller galaxies in its gravity field. Two of these satellites can be seen in the immediate neighborhood of M31. The first satellite is an eighth magnitude elliptical dwarf galaxy called Messier 32. It looks like an inconspicuous out-of-focus star, just 21' south of M31's nucleus. Messier 32 appears to touch the rim of M31's halo, as shown in the sketch (see Fig. 13.8). M32 is so small that you will not note its presence unless you really look for it. This elliptical dwarf galaxy is about 6,500 ly large. Messier 32 was likely a much bigger galaxy before M31 robbed it of many of its stars, gas and dust. What's left over of M31 is the bulge of a former galaxy.

The second companion of M31 is a lot harder to see. It is the dwarf galaxy called Messier 110. It has the same eighth magnitude brightness as M32, but it is at least twice as large. M110 is what we call an object with a very low surface brightness. It looks like a faint isolated cloud a half degree NW of M31's central bulge. This dwarf galaxy is about 11' large and appears slightly oval. If you have trouble seeing M110's weak glow, try to gently wobble your binoculars while you take advantage of your averted vision. The moving stimulus will enhance the sensitivity of your eyes' retinas. (The same technique can of course be applied to tracing out the extremities of M32's halo).

With a size of 16,000 ly, Messier 110 is the largest accompanying galaxy of M31. Both M32 and M110 consist mainly of evolved stars. Satellites like these are constantly being robbed of mass by their mother galaxy. Therefore they lack the resources to form new generations of stars. It is possible that the globular clusters, found in many galaxies, are nothing more than the shattered pieces of swallowed up galaxies. The Milky Way also has its fair number of satellites, of which the Large and Small Magellanic Clouds are the brightest specimen. Unfortunately, the Magellanic Clouds remain hidden below the horizon for mid-northern observers.

The Andromeda Galaxy can be observed from early autumn until late winter. Together with its satellites, the Andromeda system is an excellent object to test your observing conditions. The size of M31's halo together with the appearance of M110 can be used to rate the quality of the night sky.

Messier 33

Now that you had the opportunity to study the bright Andromeda Galaxy, let us raise the bar and face a more challenging galaxy in the neighboring constellation of Triangulum. Our subject is Messier 33, often called the Pinwheel Galaxy.

Messier 33 has two very promising characteristics. It is a bright mag 5.7 object, and it is twice as large as the full Moon. So you might expect that observing M33 is a walk in the park. The contrary is true. This galaxy is so huge that its light is spread over an area too large to be easily seen in a light-polluted sky. Messier 33 has proven to be one of the most frustrating Messier objects for amateur astronomers to go after. Many observers have sought to find M33, but only a few have confirmed its presence in the eyepiece. The galaxy can easily eluded you. The problem may be that you are using the wrong equipment. A powerful telescope may simply be too much to separate the weak glow of Messier 33 from the gray backdrop. So be sure to use the lowest possible power of your telescope. It will condense the galaxy's halo into a smaller and brighter form. The aperture of your equipment does not play a role, but light pollution does. You may never see Messier 33 with the naked eye from your own suburban backyard. However from a dark location in a mountain region, you can hope to enjoy the naked eye view of both the Andromeda Galaxy and Messier 33.

Locating the position of M33 in the sky is a lot easier than actually seeing it. The Pinwheel Galaxy can be found about halfway between third magnitude α Trianguli and second magnitude β Andromedae (see Fig. 13.6). Messier 33 lies at about the same distance from us as the Andromeda Galaxy. Its angular distance of 15° to the Andromeda system equals a true separation of roughly 1 million light-years. Messier 33 is the third spiral galaxy that belongs to the Local Group. The other two are the Andromeda Galaxy and the Milky Way.

Messier 33 has a true size of 60,000 ly, the third largest galaxy of the Local Group. Most of the brightness of the galaxy comes from the loose spiral arms, because M33 does not exhibit a central bulge. The spiral arms, in which numerous star formation regions are active, can be followed right into the nucleus of the system. The lack of a bright bulge is what makes the visual detection of the Triangulum Galaxy so difficult. Many galaxies can be seen in binoculars just because of their relative bright bulges.

Messier 33 does not contain a supermassive black hole in its core. This might mean that the Pinwheel is a relative young system, which hasn't yet had the opportunity to swallow much mass from smaller galaxies?

What can you expect to see from the Pinwheel Galaxy in your binoculars or small telescope? Unfortunately, unless you observe from a dark site, M33 won't reveal much details. The Pinwheel Galaxy may show up in your binoculars as a weak amorphous shimmer. Gently rocking the instrument does help to increase the contrast of Messier 33's contours against the background glow. From a dark site, a pair of 15 × 70 binoculars revealed the oval shape of the galaxy's halo, whose brightness slightly increases towards its center. With patience, you may detect the non-stellar nucleus of the Pinwheel Galaxy as well. However, if you use a telescope with too much power, you'll probably look through the galaxy without noting its presence at all (Fig. 13.9).

NGC 752

A large nebula at 8° NE of M33 fills the space between Triangulum and Andromeda. When you aim your binoculars at that spot, you should discover the mag 5.7 galactic cluster NGC 752 (Fig. 13.6).

How come Charles Messier did not include this 50′ large gem in his deep-sky catalog? Could it be that Messier's instruments magnified so much that he simply overlooked the loose cluster? NGC 752 is a perfect binocular object. Just aim your binoculars at 3rd magnitude β Trianguli and move three and a half degrees to the NW. At low power, NGC 752 resolves completely in a collection of mag 8 and fainter stars. A pair of 8 × 56 binoculars

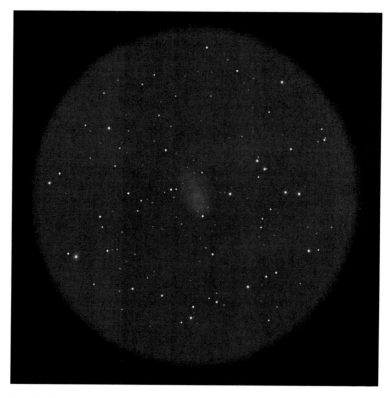

Fig. 13.9 Messier 33, observed with a pair of 15×70 binoculars

shows 16 stars in a hazy spot, while a 15×70 pair doubles that figure. With a small telescope, all of the 80 members of NGC 752 can be seen (Fig. 13.10).

The sparse cluster lies at 1,500 ly away. Its lucida is a G7 yellow giant. Because NGC 752 is a very old cluster with an age of 1.4 billion years, all its massive stars are already burned up. That a loose cluster can exist for such a long time is maybe due to the fact that NGC 752 has worked itself away from the galactic plane where the disk stars exhibit their strongest gravitational pull.

Fig. 13.10 NGC752, observed with a pair of 15×70 binoculars

Southwest of NGC 752 lies the bright double star, 56 Andromeda. It is easily split in a pair of binoculars. Both companions (of mag 5.7 and mag 5.9) are orange K-type stars. See if you can detect their color.

After you've thoroughly observed NGC 752 in the eyepieces, why not attempt to notice its glow with the naked eye? If so, compare NGC 752 with the famous Andromeda Galaxy. Bear in mind that Messier 31 (and Messier 33) lies about 1,600 times further away in space than NGC 752.

DECEMBER

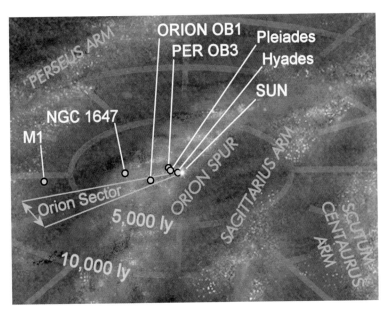

Fig. 14.1 The December objects all lie in the direction of the anti-center of the Milky Way

December announces a new season: winter. It also features the winter solstice in the northern hemisphere. By the end of the month, the days are growing longer again. During this month, the Sun passes in front of the galactic bulge of the Milky Way. As a result, our night-time window lies in the direction of the anti-center of the Milky Way (see Fig. 3.1). The anti-center is the opposite direction of the galactic center. It is a point on the galactic equator near the border between Taurus and Gemini. Messier 35 lies only a few degrees away from the galactic anti-center. December is

R. De Laet, *The Casual Sky Observer's Guide*, Astronomer's Pocket Field Guide, DOI 10.1007/978-1-4614-0595-5_14, © Springer Science+Business Media, LLC 2012

the least comfortable month to go stargazing, but this month's celestials treasures are among the most famous objects in the sky. So dress warm and enjoy the tour. We will focus our attention on the constellations of Perseus, Taurus (the Bull) and Orion (Fig. 14.1).

The Alpha Persei Association

The conspicuous constellation of Perseus passes through the zenith on December evenings. We can make use of this occasion to take a good look at Perseus' brightest star, α Persei or Mirphak. The mag 1.8 Mirphak is featured in the finder chart of Fig. 13.6. Perseus' brightest star is a spectacular F5 supergiant. It has about 8 solar masses and 62 solar radii. Mirphak is about 600 ly away and shines with the luminosity of 5,000 Suns. The star has evolved from a hot B-star to a much cooler F-star with a surface temperature of 6,180 K. If we could put our Sun next to Mirphak, then we would see it as a mag 11 star.

On a dark and moonless night, Mirphak shines from within a nebulous field. When we aim our binoculars on Mirphak, we see one of the finest star fields ever. The field between Mirphak and δ Persei is occupied by a densely packed star stream. Even the smallest binoculars have no trouble showing these curving star streams. The true nature of this star field was discovered in the twentieth century by measuring the motion of the individual stars around α Persei (Fig. 14.2).

The majority of these stars share the same proper motion true space, which led to the system's name: the Alpha Persei Association, or Perseus OB3. In 1915, Philibert Jacques Melotte, a British astronomer with Belgian roots, included the widespread cluster in his catalog of deep-sky objects as the twentieth entry. Melotte 20 is a 50-million-year-young association at a distance of 620 ly. It consists of many hot B-stars, of which 3rd magnitude δ Persei is the brightest specimen. Delta Persei is believed to be 50 million years old. This B5 star has about 6.5 solar masses and 10 solar radii. Its luminosity equals 3,400 Suns. Delta Persei is at the end of its main sequence lifetime. The star has used up all the hydrogen in its core and is moving towards the realm of the red giants.

Fig. 14.2 Melotte 20, observed with a pair of 8×40 binoculars. The two brightest stars are Mirphak and δ Persei (*left*). The fov measures 8.2°

Melotte 20 is one of the most pleasing objects for binocular observers. Its curving stellar stream has also led to its amusing nickname, the Saxophone cluster. See if you can recognize this Belgian woodwind instrument in the three-and-a-half-degree-wide association. We should not forget that Mel20 is a beautiful and bright naked-eye object as well. Its integrated magnitude is about mag 1. Seven stars of Melotte 20 shine brighter than fifth magnitude. How many stars from the Alpha Persei Association can you resolve with the naked eye?

The Hyades

South of Auriga and Perseus lies the constellations of Taurus. Taurus depicts the face and the horns of a celestial bull. Many of Taurus' stars are physically related. The V-shaped face of the bull is a true cluster, called the Hyades or Melotte 25. We inherited the name of the Hyades from Greek mythology. The constellation's brightest star is the red giant, Aldebaran. It conveniently represents the bull's fierce right eye. However, Aldebaran is not a true member of the Hyades (Fig. 14.3).

The Hyades are an evolved cluster at 150 ly away. After the Ursa Major Moving Cluster (see earlier), the Hyades are the second closest cluster to our Solar System. Its age is about 625 million years. The 4.5° wide V-shape represents only the core of the cluster. It corresponds with a real size of 12 ly. The true physical extension of the Hyades spans a total of 10° of our sky. About 380 stars are believed to belong to this cluster. With the naked eye, several of its double stars can be resolved. Theta Tauri forms a nice 5'-wide

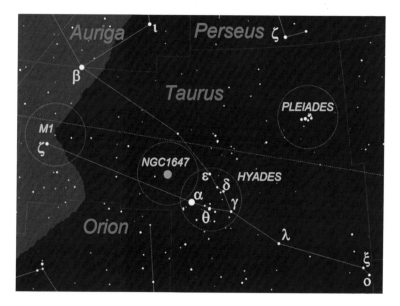

Fig. 14.3 Finder chart for the December objects

Fig. 14.4 The Hyades, observed with a pair of 8 × 40 binoculars. The bright orange Aldebaran is a foreground star

pair, easily separated with the naked eye. Delta Tauri forms a beautiful triple star. Many of the cluster's stars shine with an orange hue. Theta one, Delta one and Epsilon Tauri are all orange K-giants. The Hyades are a very sparse cluster. Its most massive stars have died long ago. The remaining brightest cluster members once were massive A-stars.

The intense orange Aldebaran (α Tauri) seems to belong to the Hyades as well. In actuality it is a foreground star. The name Aldebaran comes from the Arabic and means 'the Follower.' Aldebaran appears to follow the

Pleiades, a cluster we will discuss later in this chapter. Alpha Tauri is a mag 0.8 K-giant at only 68 ly away. Its mass is estimated at 1.7 solar masses, and its radius is about 44 solar radii. When we look at Aldebaran, we can imagine the fate our Sun will face as it, too, will become such a luminous giant. When we picture the Sun next to Aldebaran, it would shine at mag 6.8. At the distance of the Hyades, our day star would shine like an 8th magnitude star (Fig. 14.4).

The Hyades really can't handle a high power view. You should try your widest field of view instrument to observe them. You might prefer to observe the Hyades with 8×40 binoculars, as this pair offers a nice wide field of view. But even without any optical aid, the Hyades remain a superb object in the winter sky. Because the cluster is located near the ecliptic, it is often visited by one of the planets. Sometimes the Moon passes in front of the Hyades, too. These occultations are spectacular events when observed with a small pair of binoculars.

Messier 45

Taurus harbors another famous cluster, called the Pleiades. This radiant cluster of first magnitude, which lies 13° NW of Aldebaran, was known since remote antiquity (Fig. 14.3). We still use the name that comes from Greek mythology, the Pleiades, or the Seven Sisters. Charles Messier included the object in his catalog as well. With the naked eye, Messier 45 does look like an impressive one-degree-wide gathering of mag 2 to mag 4 stars. The Pleiades are so bright that even a light-polluted site will be sufficient to show them without optical aid. But their true majesty is displayed in the low power field of a pair of binoculars. Notice how the Seven Sisters radiate with a bluish hint. Their specific color becomes more apparent when you compare the Pleiades with the tone of the neighboring field stars. The Pleiades are all hot B-stars. Their age is estimated at 100 million years. The Pleiades Star Cluster is the fourth nearest star cluster, at a distance of 425 ly. The Seven Sisters form the 7 ly wide core of a 70 ly large system consisting of over 1,000 members.

We can compare the asterism formed by the Pleiades to a Little Dipper. Depending on your sky conditions, you can resolve six to nine (or even more) stars in the Pleiades without optical aid. The cluster's lucida is the mag 2.9 Alcyone, a giant B-star with six times the solar mass and ten times the solar radius. It is the leftmost star of the bowl of the Little Dipper. The next brightest member is the mag 3.6 Atlas, which forms the handle, east of this little bowl.

With a pair of binoculars, three small starlets can be detected 2′ west of Alcyone. Beware that Alcyone's luster will make this trio very inconspicuous. Notice how a fine chain of mag 7 and fainter stars runs from Alcyone to the south.

About 15′ west of this star chain shines the southernmost Pleiade, called Merope. Merope is interesting from a deep-sky observer's perspective, as it has given its name to an ethereal reflection nebula, the Merope Nebula. The Pleiades are indeed surrounded by nebulosity. The cluster seems to travel through a nebulous stretch of the Taurus Dark Cloud Complex. The Merope Nebula is the brightest part of nebulosity, in which all the bowl stars of this Little Dipper are involved. Be aware that this nebulosity is an extremely faint feature, like the breath on a mirror. Not the aperture of you equipment but the darkness of your sky determines the detection of this insubstantial reflection nebula. The sketch (see Fig. 14.5) shows what you can expect to see from a very dark location. The Merope Nebula looks like a delicate cone of light, south of Merope. A magnification of 15 times or more is mandatory to notice a small trail of the Merope Nebula just north of Merope.

Do not mistake the possible dew on your optics for the Pleiades reflection nebula. If you are convinced that you can detect the Pleiades reflection nebula, check the glare around Atlas. On the darkest of nights, the bowl stars exhibit a larger halo due to the reflection nebula than Atlas. If Atlas also shows a large halo, then your optics are suffering from dew.

Perhaps the most exciting view is the naked-eye comparison between the Pleiades and the Hyades. When you imagine that the Pleiades are three times further away than the Hyades, the Taurus constellation suddenly turns into a three-dimensional piece of sky.

Fig. 14.5 The Pleiades, observed with a pair of 15×70 binoculars

NGC 1647

With such outstanding showpieces such as the Hyades and the Pleiades so closely together in the sky, observers often neglect the lesser known Taurus objects. One of them is the fine binocular cluster called NGC 1647. Without the Pleiades or the Hyades, NGC 1647 would have been one the highlights of Taurus.

NGC 1647 is a 100-million-year-old galactic cluster at a distance of 1,800 ly. Its apparent diameter of 40' corresponds with a true size of 23 ly. The clusters combined brightness of mag 6.4 is nothing in comparison

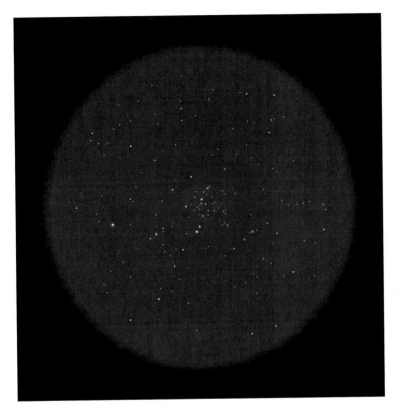

Fig. 14.6 NGC1647, observed with a pair of 15×70 binoculars

with the luster of the previously mentioned objects. Nevertheless, NGC 1647 definitely deserves a good look through your binoculars.

NGC 1647 is an easy find at 4° NE of Aldebaran (see Fig. 14.3). The majority of the cluster's 200 or so stars remain unresolved in a small pair of binoculars. A pair of 8×56 can show 12 stars embedded in softly glowing haze. With a pair of 15×70 binoculars, you could resolve over 30 cluster members. A nice close pair of mag 9 stars appears near the center of the large cluster (Fig. 14.6).

NGC 1647 and the Hyades appear to be close neighbors in the sky. However, in reality, they are located on far extremities of the Orion Spur (see Fig. 14.1). The Hyades lie near the inside edge of the Orion Spur, like our Solar System does. NGC 1647, however, lies 12 times more distant than the Hyades. It lies on the outside edge of our spiral arm. When switching our gaze from NGC 1647 to the Hyades, we practically bridge the cross-section of our spiral arm.

Though the mag 6.4 NGC 1647 has been reported as a naked-eye object, not many are able to see it without optical aid.

Messier 1

Our next destination represents perhaps the most spectacular event that can be seen with a small pair of binoculars – a supernova remnant!

When a massive star nears its life's end, its core runs out of fuel. The hot dense core no longer produces the energy or the pressure to withstand the gravity of the star's envelope. The star's core suddenly collapses into a tiny ball of degenerated matter. This ultimate nuclear reaction together with the rebound of the infalling matter on the degenerated core releases so much energy that the star literally explodes. Astronomers call this explosion a supernova.

When a star produces such a catastrophic blast, it releases as much energy in a single moment as a hundred Suns produce during their entire life span. A supernova becomes brighter than a whole galaxy. Seen from afar, it looks like the birth of a new star (in Latin, stella nova). Supernovae are important processes as they are responsible for the chemical enrichment of interstellar matter. All the chemical elements that we find in our Solar System are produced and ejected by the nuclear explosions of supernovae. This ejecta contains microscopic grains that astronomers like to call dust. So we are all really made up of stardust, as all the elements that are critical for life on Earth are manufactured by older generations of stars. Supernovae are at the same time extremely hazardous for life in their immediate environment and yet they are indispensable for generations of life to come.

Only a few supernovae have been recorded through human history. One such written record can be found in the Chinese annals of the Sung

Dynasty. On July 4, 1054, a very bright "guest star" was discovered next to the crescent Moon in the constellation of Taurus. The Chinese astronomers could see the nova during daytime until the 27th of July of the same year. At night, the star could be seen until the17th of April 1056. This new star was probably brighter than Venus, and possibly brighter than the full Moon. The event must have been the 'hot' item of the month throughout the whole world. Imagine how the night sky was lit up by this new guest star, adjacent to the crescent Moon. Imagine how people everywhere interrupted their normal daily cycle to admire the brilliant guest star. And nobody could explain what had happened.

Nearly 677 years later, in 1731, the English amateur astronomer John Bevis discovered a small nebula in Taurus. At that time, he did not realize that he had discovered the supernova remnant of the 1054 supernova. In 1758 Charles Messier, while looking for comet Halley, found the object independently. At first Messier thought that he had found comet Halley. He soon realized that this comet-like patch did not have proper motion. This event had two important consequences. It caused Messier to create his Catalogue of Nebulae and Star Clusters, and it motivated Messier to start hunting for comets with telescopes. The nickname 'Crab Nebula' had its origin in 1844, when William Parsons made a detailed sketch of the resolved filaments within M1. Not until 1921 was the Crab Nebula linked with the supernova of 1054.

Today, almost 1000 years after the catastrophic explosion of 1054, the afterglow can still be seen, even with small instruments. Look for a weak 5' wide nebula at one degree NW of Zeta Tauri, the tip of the southern horn of Taurus (Fig. 14.3). This little puff of smoke is the only fragile recollection of a violent explosion, 6,200 ly away, which was bright enough to read a newspaper at night. M1 does not look particularly bright through a pair of binoculars. However, the supernova remnant is still brighter than our Sun, which would shine at mag 16 when placed next to the Crab Nebula. This means that after almost a millennium, the wreckage of this hot and massive O-star is still 1,500 times brighter than our Sun. The Crab Nebula is still expanding. Its current size equals roughly 10 ly (Fig. 14.7).

Buried deep inside the debris of the supernova remnant, lies the pulsing heart of a long-gone star. The collapsed dense core has produced a fast rotating pulsar, an exotic, super-massive body that measures only 20 km

Fig. 14.7 The Crab Nebula, observed with a pair of 15×70 binoculars. The bright star is ζ Tauri

across. Its gravity at the surface is so strong that the pulsar's light mainly escapes through its sweeping magnetic poles. Our Solar System happens to lie in the line of sight of the tilted magnetic poles. As a result, the beam of the Crab Nebula pulsar sweeps over our Solar System 30 times per second, like a lighthouse sweeps its beam over the coastline.

We are fortunate to witness the expansion of this supernova remnant before it vanishes forever. The glow that we observe from the Crab Nebula is caused by the radiation of its pulsar. Supernova remnants are, like planetary nebula, very short-lived phases of a star's life. We are also fortunate that our Sun has always been at a safe distance from such supernovae. The blast of the 1,054 supernovae is believed to have had the ability to destroy all life within a radius of 60 ly of the exploding star.

The Orion OB1 Association

Our final destination is the constellation of Orion. The binocular show-pieces of Orion have already been treated in Chap. 3. For now, let us enjoy the naked-eye impression of mighty Orion as shown in Fig. 14.8. The majority of the constellations are made up of stars that just happen to lie in the same line of sight. They have no astrophysical relationship whatsoever with each other. One of the exceptions is indeed Orion. Here we see not just an imagined pattern in the sky, but a real stellar association.

Fig. 14.8 Sketch of Orion during twilight on a spring evening. The Hyades are visible in the upper right. Sirius shines brightly from between the trees at the left

A stellar association is a very loose gathering of stars that share the same origin. These stars are no longer bound by gravity, but they still move together through space. They are often called OB associations because these young stellar complexes contain scores of massive O- and B-stars that formed within the same giant molecular cloud.

The eye-catching Belt stars, ζ, ε, δ, together with Saiph, Rigel, λ and ι, all belong to the Orion OB1 association (see Fig. 3.8). They are the hot O- and B-stars emerging from an immense molecular cloud, the Orion Molecular Cloud Complex, which envelops a substantial part of the constellation. The Orion complex represents a considerable stretch of our local spiral arm, the Orion Spur. Here we see the creation of stars and clusters in full progress.

With Orion OB1 in our own backyard at 1,200 ly, we are privileged to witness the activity of an extremely young and giant stellar nursery from nearby. The Orion Complex is several hundreds of light-years deep. Rigel lies at 800 ly, Saiph at 700 ly, lambda Orionis at 1,100 ly, the Belt stars at 1,200 ly and Iota Orionis at 1,300 ly. Whether Betelgeuse is a true member of Orion OB1 or not is still being debated. Betelgeuse is an M2 red supergiant at 650 ly away.

There are no red giants found in the young Orion OB1 association. Betelgeuse, which is only a few million years old, could still be among the first stars that formed here. What can be said with certainty is that Betelgeuse, with his 18 or so solar masses, is running out of fuel. This behemoth, which is about 1,000 times larger than our Sun, is due to go supernova within the next few thousand years. Betelgeuse will then brighten from its current mag 0.42 to −12 and become visible in broad daylight, for several weeks. Supernovae can produce extremely hazardous radiation, but Earth is luckily at a safe distance.

Bellatrix is a blue giant B2 star at a distance of 240 ly. It is probably too close to belong to the Orion OB1 Association.

Orion is lean in open clusters and yet it lies near the galactic equator. The reason is that the dense molecular clouds absorb the light from stars within and beyond. Within a few tens of millions of years, when the remains of the molecular cloud have dispersed, the young clusters will be freed from their dark molecular cocoons, and Orion will become a more transparent stretch of the Milky Way.

Sit back and admire the naked-eye view of Orion, the constellation that has inspired so many cultures since ancient times. Imagine how many generations of the human species have admired Orion before us, and how they wondered about the true nature of this mighty constellation. Our ancestors have always tried to understand the universe. They associated Orion with a hero, a shepherd, a hunter or even a warrior.

Our ancestors uplifted Orion to a divine status. Modern humans still lesser appeal to the gods to explain the structure of our Universe. Nevertheless, whether you are a faithful being or not, it doesn't change a thing to the fact that the Creation is presently underway in this wondrous part of the heavens which our ancestors admired as their hero, Orion.

CHAPTER 15

TAKING IT FURTHER

Now that you have completed all the tours, you could presume that you have seen it all. But the opposite is true. I advice you to repeat all the tours over and over again. It will feel like revisiting old friends. Many targets will reveal new details. They will look brighter and more familiar to you. Some of then will become your favourite objects to show to others. Perhaps you will spend your vacation on a distant location with a darker sky. It could offer you a better opportunity to go stargazing.

This book has only treated a few of the brighter highlights in the sky. There are so many more deep-sky objects to observe. If you want to take it further, look at the suggested books and websites below. And why not meet fellow stargazers for a group star party gathering? Observing in group is so much more enjoyable.

References

Suggested Websites

With the ever growing popularity of the Internet, amateur astronomers can easily share their experiences in cyberspace. Follow a few of these links to interesting websites and discussion boards.

http://apod.nasa.gov

http://www.asod.info

http://www.cloudynights.com

R. De Laet, *The Casual Sky Observer's Guide*, Astronomer's Pocket Field Guide,
DOI 10.1007/978-1-4614-0595-5_15, © Springer Science+Business Media, LLC 2012

http://www.darksky.org

http://www.heavens-above.com/

http://www.iau.org

http://www.spaceweather.com

http://www.visualdeepsky.org

http://www.wikipedia.org

Suggested Reading

Crossen, Craig, and Gerals Rhemann. *Sky Vistas*. Springer Wien New York. (2004)

Crossen, Craig, and Will Tirion (2008). *Binocular Astronomy*. Willmann-Bell. (2009)

French, Sue (2005). *Celestial Sampler*. Sky Publishing. (2005)

Handy, Richard. et al. *Astronomical Sketching*. Springer (2007)

Harrington, Philip S. *Touring the Universe through Binoculars*. John Wiley & Sons. (1990)

Kaler, James B. *The Cambridge Encyclopedia of Stars*. Cambridge University Press. (2006)

Luginbuhl, Christian B., and Brian A. Skiff. *Observing Handbook and Catalogue of Deep-Sky Objects*. Cambridge University Press. (1998)

O'Meara, Stephen James. *The Messier Objects*. Cambridge University Press. (2001)

Suggested Software

Cartes du Ciel / Skychart, by Patrick Chevalley, freeware

http://www.ap-i.net/skychart/start

Where is M13, by Bill Tschumy, freeware

http://www.thinkastronomy.com/M13/index.html

GLOSSARY

Anti-center   The point in the sky opposite of the galactic center.

Active galaxy   A galaxy of which the center has a much higher than normal luminosity.

Altitude   The angle between an astronomical object and the observer's horizon, also called the elevation.

Aperture   The diameter of the objective lens or mirror of a telescope.

Atmospheric extinction   The amount of light that is absorbed by the air we look through. Stars near the horizon look dimmer than stars near the zenith because we look through a much thicker mass of air. Therefore astronomical objects are best observed when they appear high in the sky.

Asteroid   A small rocky-icy or metallic body that orbits the Sun.

Asterism   A striking pattern of stars, most of which are not physically related.

Averted vision   A technique used to observe faint objects, most valuable in low light conditions. Instead of looking directly at the object, the observer looks off to the side. As such, the object is projected on the more sensitive area of the observer's retina.

Binary star   A star system consisting of two stars that orbit a common center of mass.

Black hole   A dense mass in space with such a large gravitational field that nothing can escape from it, not even light. Black holes are thus invisible. Their presence, however, can be noted by their interaction with their surroundings.

Blue giant   A giant star with a spectral type of O or B. It is a massive and very luminous star that has exhausted the hydrogen in its core. When this happens, the star leaves the main sequence and expands into an object classified as a giant star.

Blue straggler   Stars in clusters are believed to be formed around the same time. When the cluster ages, its most massive (blue and luminous) stars are the first to become giants and to die. Over time, the cluster becomes redder.

R. De Laet, *The Casual Sky Observer's Guide*, Astronomer's Pocket Field Guide, DOI 10.1007/978-1-4614-0595-5, © Springer Science+Business Media, LLC 2012

A blue straggler is a main sequence star that appears bluer and more luminous than the age of the cluster it is in can allow. A possible explanation is that blue stragglers are mergers of individual stars.

Brown dwarf   Star with a mass lower than 0.08 solar mass. Brown dwarfs don't have enough mass to fuse hydrogen. They are much cooler and dimmer than main sequence stars.

Cluster   A star cluster is a group of stars. There are two types of star clusters: globular clusters and open clusters. Globular clusters are tight groups that are firmly bound by gravity. Globular clusters can be found in the halo surrounding our Milky Way. Open clusters (or galactic clusters) are loose collections of stars. They are formed in the galactic disk of the Milky Way. Open clusters disintegrate over time due to interactions with other stars and molecular clouds in the galactic disk. Our Sun, which travels in solitary through the galactic disk, is believed to have been born in an open cluster.

Coating   A coating is used in an optical instrument to enhance the transmission of the captured light. Uncoated lenses reflect too much of the infalling light. Modern optics with multiple layers of broadband antireflective coatings on all the glass to air surfaces are labeled 'fully multicoated' (FMT).

Comet   A small icy body that orbits the Sun. When the comet comes close enough to the Sun, streams of vapor and dust start to come out of the comet's nucleus.

Conjunction   When two or more celestial objects appear in the same line of sight, then they are in conjunction.

Constellation   A group of stars that form a pattern in the sky. There are 88 official constellations. A *circumpolar constellation* is a constellation that never sets or rises, as viewed from a given latitude on Earth. It is visible all night on every night of the year. A *seasonal constellation* on the other hand is only visible in a particular period of the year.

Culmination   The maximum altitude of an object, which occurs when it crosses the observer's meridian.

Deep-sky object (DSO)   An astronomical object beyond the Solar System. Typical DSO's are star clusters, nebulae and galaxies.

Direct vision   The technique of looking direct at an object. For our direct vision, we make use of a small part of our retina, called the fovea. Direct vision offers sharp, colorful vision and works best in bright light.

Double star   A true binary star system or two unrelated stars that happen to lie in the same line of sight.

Ecliptic   The apparent path of the Sun against the background stars during the year. The Moon as well as the planets travel very close to the ecliptic, too.

Exit pupil   The plane of the virtual aperture behind the eyepiece of a telescope where the image of the observed object is projected. The observer's eye must be aligned with the exit pupil. The diameter of the exit pupil equals the aperture of the telescope divided by the magnification.

Eyepiece   The lens of a telescope through which the observer looks. The eyepiece determines the magnification of the telescope. Most telescopes have exchangeable eyepieces. Binoculars have fixed eyepieces.

Eye relief   The distance between the eyepiece and the exit pupil. It is thus the distance at which the observer should hold his or her eye in order to see the full field of view. If the observer's eye is further away from the eyepiece, there will be a narrower field of view. Eyepieces with an eye relief of 15mm or larger are suitable for eyeglass wearers.

Field of view (fov)   The true field of view is the angular extension of the observed scene that is visible in the eyepiece. Binoculars offer a true fov from 8° to 3.5°. Apparent field of view is the angular size of the image presented by the eyepiece. The apparent fov is a constant measure for a given eyepiece. Eyepieces have an apparent fov ranging from 40° to over 100°. Imagine that an eyepiece has an apparent fov of 50° and a magnification of 100. When you look at the full Moon, you will see in the eyepiece an image of the Moon as large as the whole apparent fov of 50°. But the true fov is only 30 arcminutes.

Galactic center   The center of our Milky Way.

Galactic disk   The plane of spiral and lenticular galaxies. The galactic disk contains mainly gas, dust and stars. These stars are called disk stars. Our Sun is a disk star.

HII-region   Large cloud of gas in which active star formation is taking place.

Hypergiant   An unusually massive star with a mass ranging from 70 to 250 solar masses. Hypergiant stars suffer from internal instabilities and have an extremely short lifetime. They are therefore very rare.

IC catalog   *The Index Catalogue of Nebulae and Clusters of Stars.* This is a catalogue that was compiled by J. L. E. Dreyer in the 1880s and is a supplement to the NGC catalog.

in.   Abbreviation for inch. Used to measure the size of a telescope. A 3-in. telescope has an aperture of 3 inches, or approximately 75mm. A 4-in. telescope has an aperture of 100mm.

Interstellar medium   The matter (gas and dust) that exists in the space between the stars in a galaxy.

Light pollution   An excess of artificial light. Light pollution brightens the night sky and washes out the stars, making it difficult to observe them. Cities and streetlights are main sources of light pollution.

Light-year   The distance that light travels in one year, equal to about 10 trillion km or 6 trillion miles.

Limiting magnitude (lm)   The faintest apparent magnitude that can be seen.

Local Group   A group of more than 30 galaxies that includes our own galaxy.

Local Supercluster (LSC)   A cluster of galaxy groups that includes the Local Group as well as the Virgo Cluster. Superclusters are the largest structures of the universe.

Lucida   The brightest star of a cluster.

Luminosity   The amount of energy that an astronomical object radiates per unit of time.

Magnitude (mag)   A measure for the brightness of an astronomical object. The magnitude scale was invented by the Greek astronomer Hipparchus. He divided the visible stars in six categories. The brightest stars were said to be of first magnitude, the faintest stars of the sixth magnitude. The *apparent magnitude* is the brightness of an object as seen from an observer on Earth. The *absolute magnitude* equals the apparent luminosity of an object imagined at a standard distance of 32.6 l-y. The apparent magnitude of our Sun is −26.74. It is the brightest object in the sky. The apparent magnitude of the next brightest star, Sirius, is −1.46. The absolute magnitude of our Sun is 4.8, while the absolute magnitude of Sirius is 1.42. Thus Sirius is more luminous than our Sun.

Main sequence   The generic term for stars that create energy in their cores from fusing hydrogen into helium. The main sequence phase comes right after the birth of a star. It is the most stable period in a star's life. Our Sun is halfway through its main sequence phase. Main sequence stars are also called dwarfs.

Meridian   An imaginary great circle in the sky that passes from the north point on the horizon through the celestial pole, up to the zenith, and through the south point on the horizon. A star reaches its culmination when it crosses the meridian. Meridian comes from the Latin word for midday.

Messier object   A deep-sky object from the catalog of the French astronomer Charles Messier.

Metallicity   The proportion of chemical elements other than hydrogen and helium present in an astronomical object. Older stars have lower metallicities than younger stars, because younger stars formed in a 'metal'-rich environment.

Meteor   The visible trace of a meteoroid as it enters the atmosphere, also misnamed a shooting star.

Meteoroid   A particle of dust or a rock that moves through interplanetary space.

Milky Way   Our home galaxy.

Multiple star   A star system consisting of more than two stars that orbit a common center of mass.

Nebula   An interstellar cloud of dust and gas. *Emission nebulae* produce light. *Reflection nebulae* reflect light from nearby stars. *Dark nebulae* absorb light from stars or emission nebulae in the background.

Nelm (naked-eye limiting magnitude)   The magnitude of the faintest stars that can be seen with the naked eye. The nelm is used as a benchmark for the darkness of the sky. Light polluted skies have a poor nelm, while dark skies have a great nelm.

Neutron star   The collapsed dense core of a star that went supernova. A typical neutron star has the mass of 1.3 to 2 solar masses compressed in a sphere as small as 24 km.

NGC catalogue   *The New General Catalogue of Nebulae and Clusters of Stars* was compiled by J. L. E. Dreyer in the 1880s. The NGC catalogue is based on the visual discoveries from William and John Herschel.

Night vision   The ability to see in low light conditions. Human night vision makes use of the rods of the eye. Our night vision slowly develops when we enter the dark. After about 30 minutes in the dark, our eyes are completely dark adapted. They become even slightly more sensitive during the next 90 minutes in the dark.

Nova   A variable star that undergoes an eruption. The star becomes suddenly much brighter. It takes months for its brightness to decline again. Novae are in fact close binary stars with one component a white dwarf. The latter captures gas from its companion. When enough hydrogen is collected and heated on the surface of the white dwarf, a violent nuclear reaction turns the hydrogen into helium, which is responsible for a sudden outburst of light.

OB association   A young stellar grouping that formed within a giant molecular cloud. It contains tens of massive stars of spectral class O and B.

Planet   Literally a wandering star. Our ancestors noticed that some 'stars' wandered around in the sky. Today we know that these special stars are members of our Solar System. Planets are large bodies that orbit the Sun. They don't emit light, as stars do. We can see them because they reflect sunlight. Planets that orbit other stars are called extrasolar planets.

Planetary nebula   A shell of glowing gas ejected by a star in its death throws. When this type of nebula was first discovered, it was confused with giant gas planets, hence the name. Planetary nebulae have little in common with planets.

Pulsar   A neutron star that emits a beam of radiation along its rotating magnetic poles. When the beam hits Earth, the pulsar seems to flash like a lighthouse.

Quasar   A quasi-stellar radio source. Quasars are the most luminous objects in the cosmos. They are believed to be the active centers of massive galaxies that host a supermassive black hole. Such black holes collect infalling material in a disk around the central body. The compression of the material in the disk is responsible for the radiation.

Red giant   When a main sequence star (with a mass ranging from 0.5 solar masses to 6 solar masses) has consumed all the hydrogen in its core, the core starts to contract. This contraction heats up the star's shell outside the core, where hydrogen fusion commences. The released energy increases the star's luminosity greatly. The star begins to expand and enters the realm of the red giants.

Red dwarf   A small star with a mass lower than 0.4 solar masses. Red dwarfs produce little energy and have less than a tenth of the Sun's luminosity. They are so energy efficient that are believed to have a longer lifespan than the age of the universe. This means that the first formed red dwarfs are still around.

Runaway star   A star that is moving through space with an unusually high velocity. Runaway stars are believed to have left a multiple star system after a supernova explosion.

Satellite   A naturally occurring object in order around another or an artificial object placed into orbit.

Seeing   The atmospheric turbulence that cause the twinkling of the stars. When the seeing is good, detailed observations are possible. When the seeing is bad, stars seem to dance around when looked at through a telescope.

Solar System   The Sun and all the bodies held in its gravitational field. Others stars also have their own solar systems.

Star   A hot, luminous ball of mostly hydrogen and helium. A binary star consists of a pair of stars that orbit around a common center of mass. A multiple-star system consists of a small number of stars that orbit each other.

Starburst   A region of a galaxy with an abnormally high rate of star formation.

Supergiant   A star with a mass ranging from 10 to 70 solar masses. Supergiants are very luminous and have a rather short lifespan.

Supernova   A violent stellar explosion in which most of the star's mass is blown away. Supernovae can be as bright as a whole galaxy.

Supernova remnant   The debris cloud left behind by a star that went supernova.

Transparency   A measure for the clarity of the atmosphere. When the air is polluted with vapor, dust or smog, the transparency is bad. There is more extinction and light scatter.

Variable star   A star in which the apparent magnitude changes over time. This can be due to the star's luminosity, or to the transparency of the space between Earth and the star.

White dwarf   Also called a degenerate dwarf. This is the final state of a star that lacks the mass to become a supernova. Once such stars have exhausted their core hydrogen supply, they become red giants. At the end of their red giant phase, these stars shed their outer layers in the form of a planetary nebula. What is left behind is the unveiled high density core of a former star. Its substance is inert. The white dwarf's fate is to radiate away its energy and cool down. Because white dwarfs typically have the mass of our Sun and the size of Earth, they need a very long time to cool down. Our universe is still too young to have any cooled down white dwarfs.

Wolf-Rayet star   Evolved massive star that continually ejects large amounts of mass into space.

Zenith   The point in the sky directly above the observer.

Index